METHODS AND APPLICATIONS OF NONLINEAR DYNAMICS

Other ACIF Publications in this Series

1. *Stochastic Processes Applied to Physics and Other Related Fields*, edited by B. Gómez, S. M. Moore, A. M. Rodriguez Vargas, and A. Rueda (1983).

2. *Workshop on the Search for Gravitational Waves*, edited by E. Posada and G. Violini (1983).

3. *Non Conventional Energy Sources*, edited by G. Furlan, H. Rodríguez, and G. Violini (1984).

4. *Galaxies, Quasars and Cosmology*, edited by L. Z. Fang and R. Ruffini (1985).

5. *Reactor Physics for Developing Countries and Nuclear Spectroscopy Research*, edited by K. Lieb and G. Medrano (1986).

6. *Proceedings of the First Latin American School on Biophysics*, edited by R. Fayad, A. M. Rodríguez-Vargas, and G. Violini (1987).

METHODS AND APPLICATIONS OF NONLINEAR DYNAMICS

ACIF Series – Volume 7

Medellín, Colombia
September 1-5, 1986

Editor
A. W. SÁENZ
Naval Research Laboratory, Washington, D.C. 20375, U.S.A.
and
Catholic University, Washington, D.C. 20064, U.S.A.

World Scientific
Singapore • New Jersey • Hong Kong

Published by

World Scientific Publishing Co. Pte. Ltd.
5 Toh Tuck Link, Singapore 596224
USA office: 27 Warren Street, Suite 401-402, Hackensack, NJ 07601
UK office: 57 Shelton Street, Covent Garden, London WC2H 9HE

British Library Cataloguing-in-Publication Data
A catalogue record for this book is available from the British Library.

METHODS AND APPLICATIONS OF NONLINEAR DYNAMICS

ISBN-13 978-9971-50-333-8
ISBN-10 9971-50-333-6

EDITOR'S FOREWORD

The present volume on *Methods and Applications of Nonlinear Dynamics* arose mainly from lectures given at the First International Course on Nonlinear Dynamics, which took place in Medellín, Colombia, on 1-5 September 1986. The aims of the Course were to discuss some of the fundamental theoretical ideas of modern nonlinear dynamics and their application to selected areas of physics, and also to help the participants to bridge the gap between textbook presentations and the contemporary research literature. The lectures were intended for and delivered to a Ph.D.-level audience composed of physicists and mathematicians. They were not primarily intended for experts, but rather for scientists interested in performing experimental or theoretical research on nonlinear dynamical phenomena occurring in real physical systems.

The volume opens with the survey lectures of Prof. A. F. Rañada (Universidad Complutense, Madrid) on the "Phenomenology of Chaotic Motion". They constitute a very readable introduction to chaotic dynamics and emphasize its physical aspects. The next contribution to the volume are the lectures of Prof. G. Turchetti (University of Bologna) on "Perturbative Methods for Hamiltonian Maps", which survey classical and recent contributions to this topic. In view of the importance of Hamiltonian dynamical systems in celestial mechanics, plasma physics, and many other areas, as well as the central importance of the contributions of contemporary Italian dynamicists to this wide subject, the inclusion of these lectures is more than justified. Prof. L. Vázquez (Universidad Complutense, Madrid) emphasizes the physical, intuitive aspects of nonlinear stochastic systems in his lectures on "An Elementary Introduction to Stochastic Processes", the next major topic discussed in the volume. The latter concludes with two survey articles by Prof. N. B. Abraham (Bryn Mawr College) and Prof. A. Zettl (University of California at Berkely) on "Chaos in Optical Systems" and "Chaos in Solid State Systems", respectively. These are two extremely active and fruitful areas of experimental and theoretical research in nonlinear dynamics in which there is active interest in Colombia and other Latin Americal countries.

Space limitations prevented the inclusion of survey articles on such important topics as chaos in fluid systems, although some salient phenomena in the latter domain are considered in Prof. Rañada's contribution to this volume.

I wish to express my deep appreciation to Profs. Rañada, Turchetti, Vázquez, and Zettl for generously contributing their time and expertise to the Medellín Course and to this book, as well as to Prof. Abraham (who for reasons beyond his control was unable to be in Medellín) for his contribution to the volume. The idea of the Course originated in conversations with Dr. S. M. Moore, whom I thank for many helpful suggestions. It was sponsored by the Asociación Pro-Centro Internacional de Física (ACIF) of Bogotá, Colombia, and Dr. E. Posada F. and Prof. G. Violini, the President and Executive Secretary, respectively, of ACIF, deserve thanks for their strong support of the meeting. My lifelong friend, Dr. Alberto Vásquez R., who was then governor of the Department of Antioquia, provided generous financial support. The Course was also sponsored by the Organization of American States. The National Organizing Committee was composed of Profs. E. Alvarez, J. Mahecha, and F. Medina (University of Antioquia), and Prof. R. Castañeda (National University of Colombia), to whom sincere thanks are due for their many efforts. Last but not least, I am grateful to Miss S. E. Dixon and Mrs. S. M. Montgomery of the Naval

Research Laboratory for the care with which they typed the relevant editorial changes to create a camera-ready manuscript, and to Mrs. H. S. Oxley of NRL for her valuable help with bibliographical matters.

It goes without saying that it is hoped that this volume will be useful to Latin American scientists and students, as well as those in the United States, Europe, and elsewhere interested in nonlinear physics.

A. W. Sáenz
Naval Research Laboratory
Washington, DC 20375
and
Physics Department
Catholic University
Washington, DC 20064

CONTENTS

PHENOMENOLOGY OF CHAOTIC MOTION

ANTONIO F. RAÑADA
Departamento de Física Teórica
Universidad Complutense
28040 Madrid, Spain

CONTENTS

CHAPTER 1 CHARACTERIZATION OF CHAOTIC MOTION

1.1 INTRODUCTION.

The realization that nature is much more complex and its behavior far richer than what was thought is one of the most appealing lessons of recent physics, since because of our apparent ability to predict with great precision the evolution of physical systems, our image of the world had become too simple, poor, and even cold. With the growing awareness that traditional physics gives only a first approximation to an essentially nonlinear world, a new perspective is emerging which approaches such seemingly unrelated problems as hydrodynamical turbulence, chemical kinetics, or celestial mechanics from the same point of view. In fact, it is found every day that systems which are described by very different models have many common traits and mathematical similarities which allow a unified treatment. To such an extent that it is possible to speak of universal patterns in the behaviour of nonlinear systems.

The most noteworthy rupture with traditional thought is the understanding that it is impossible, as a matter of principle, to predict precisely the long time behaviour of many systems because of the extreme instability of the solutions of their equations of motion. This was already known to Poincaré[1], who had studied this problem in his "Méthodes nouvelles de la mécanique céleste". But physics took the promising road of quantum theory and this question was forgotten, in spite of some warnings by Einstein.[2] Many years later, in 1954, it came again to light when Kolmogorov[3] stated what is now known as the KAM theorem (after his initial and those of Arnold[4] and Moser[5] who completed the proof), which characterizes the way in which instability arises when an integrable, and therefore regular, Hamiltonian is perturbed. In spite of its great importance, it applies only to Hamiltonian systems and not to the more abundant dissipative ones. The next development happened in 1963, when the meteorologist E.N. Lorenz [6)] found completely chaotic trajectories in a very simple dissipative system of three coupled ordinary differential equations. The great relevance of his result was not recognized until a decade later, but it is widely appreciated today and is often used as a paradigmatic example.

Such complex behavior is neither due to external noise, nor to a great number of degrees of freedom, nor to quantum effects. It just happens that trajectories, which in a certain moment are very close, separate exponentially in time and intermingle with others in an erratic and violent way. To designate this phenomenon, Lorenz coined the expresion *"butterfly effect"*, thus alluding to the sudden change in the initial conditions of the atmosphere because of the unexpected beating of the wings of a butterfly.

For classical physics all systems were, in principle, predictable,this idea being at the conceptual basis of mechanicism, which had such great importance in modern thought. But in the 19th century the study of systems with many degrees of freedom forced the beginning of a new probabilistic tradition, as their detailed study was thought to be impossible because of practical reasons. These two traditions were thought to be complementary, rather than opposite, the general attitude being the following: if a system has few degrees of freedom, use the techniques of Newtonian dynamics; if many, those of the new statistical mechanics. For this point of view, complex behaviour appears only when there are simultaneously many simple elements, so that systems with few coordinates must behave simply.

However, and to the surprise of most, even systems with only two degrees of freedom exhibit the butterfly effect or, in other words, their time evolution is turbulent and chaotic. This is called deterministic chaos, since the governing equations are deterministic but have solutions exhibiting stochastic properties. This discovery is causing a very deep change in our ideas on motion and, according to many, is bringing about a conceptual revolution which can be compared to those of relativity or quantum theory. Maxwell, one of the founding fathers of statistical mechanics, glimpsed it, when he wrote (quoted by {Berry, 1978})(curly brackets denote general references):

"If, therefore, those cultivators of the physical sciences. . . . . are led in pursuit of the arcana of science to the study of the singularities and instabilities, rather than the continuities and stabilities of things, the promotion of natural knowledge may tend to remove that prejudice in favor of determinism which seems to arise from assuming that the physical science of the future is a mere magnified image of that of the past".

In the case of Hamiltonian systems, it is possible to characterize in a simple way those which are regular, that is, nonchaotic. These are the integrable systems, so called because their equations of motion can be reduced to quadratures. Systems in this class have as many independent constants of the motion with vanishing Poisson brackets as they have degrees of freedom. On the other hand, those for which this is not the case, as happens in the famous three body problem, are called non-integrable and they exhibit chaotic behavior.

In these lectures we will not deal with such systems, but with the more abundant nonconservative ones, which have a richer and more diverse behavior. There is, in particular, a very important difference concerning stability. Conservative systems, contrary to dissipative ones, cannot have asymptotically stable solutions; however,this kind of stability is essential to many of the natural world structures, as this occurs in the important case of living beings.

1.2 CHAOS IS UBIQUITOUS

Since Lorenz's paper the list of systems in which chaos has been shown to exist, either experimentally or by the numerical solution of the equations of motion, grows at an accelerated pace. We can mention the following:

- Hamiltonian systems
- Celestial mechanics (e.g., the three-body problem)
- Fluids
- Lasers

- Nonlinear optical systems
- Solid state
- Particle accelerators
- Plasmas
- Chemical reactions
- Population dynamics
- Biological systems (e.g. the heart cell, the brain cell)

Let us consider some examples.

(a) The Forced Pendulum

The pendulum is frequently presented in elementary textbooks as one of the simplest physical systems. However, it shows chaotic behaviou if it is damped and is sumbmitted to a periodic force. [7,8] This example has great interest, because the equation of motion is the same as that of a Josephson junction submitted to microwave radiation:

$$\ddot{\phi} + \frac{1}{\tau}\dot{\phi} + \omega_o^2 \sin\phi = \Gamma \cos \omega_d t, \tag{1.1}$$

where $\dot{\phi}$ is proportional to the potential difference, τ is the damping time, ω_d is the natural frequency and ω_d that of the microwave. For some values of Γ and ω_d there are chaotic solutions (see Fig. 1.1).

(b) The Sitnikov Case of the Three-Body Problem

That the three-body system is chaotic is shown very clearly in the particular case first studied by Sitnikov[9,10] (see also (Berry,1978)). As is seen in figure 1.2, two equal primaries of mass M follow elliptic orbits of excentricity ϵ, around their center of mass G in the plane π. A planetoid of negligible mass moves in their gravitatory field, along the straight line orthogonal to π at G.The equation of motion is

$$\ddot{z} = - \frac{z}{\{z^2 + \frac{1}{4}(1 - \cos 2t)^2\}^{3/2}} . \tag{1.2}$$

If the orbits are circular, ϵ=0 and the system is integrable, the motion being periodic and consisting in a nonlinear oscillation around G.

If not, the motion is aperiodic and chaotic. If t_k are the values of time when the planetoid passes through the plane π, we can define the "periods" as $\tau_k = t_{k+1} - t_k$. It can be shown then that there exists a function $S(\epsilon)$, such that $S(\epsilon)\to\infty$ if $\epsilon\to 0$, with the property that for any random sequence $a_k > S(\epsilon)$, there is an initial condition $(z(0), \dot{z}(0))$ such that $\tau_k = a_k$; in other words there are infinite sets of solutions which can be characterized by random sequences of numbers. And this is so, in spite of the apparent simplicity of the system, which has only one degree of freedom.

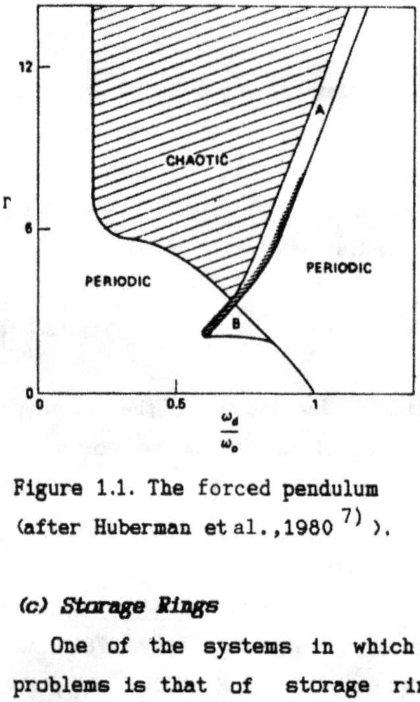

Figure 1.1. The forced pendulum (after Huberman et al., 1980 [7]).

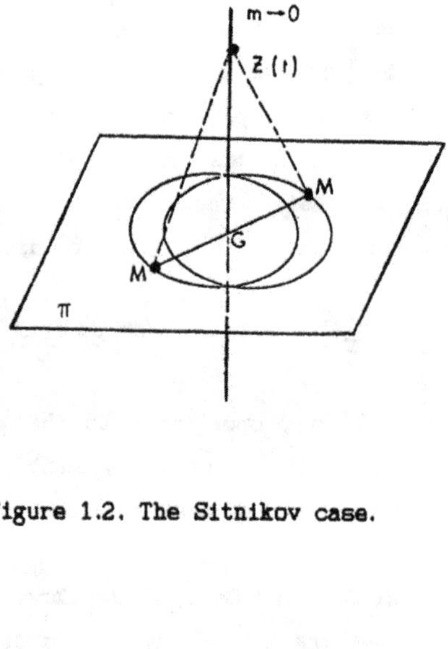

Figure 1.2. The Sitnikov case.

(c) Storage Rings

One of the systems in which chaos is posing difficult design problems is that of storage rings, in which charged particles move in electromagnetic fields ({Helleman, 1980}) . If only one of the beams is travelling along a circular path, the transverse oscillations of the protons are described by the equation

$$\frac{d^2y}{d\phi^2} = - Q^2 y,$$

where y is the normal deviation, ϕ is the angular coordinate and Q is a characteristic frequency. The collisions with a second beam are practically instantaneous and can be modelled by

$$\frac{d^2y}{d\phi^2} = - Q^2 y + B\ F(y) \sum_n \delta(\ \phi - 2\pi n)\, . \qquad (1.3)$$

Although very few of the protons collide, all suffer a strong nonlinear force BF(y). The system (1.3) cannot be solved analitycally and its numerical treatment is very difficult, since in a typical experiment, the protons undergo more revolutions than those of by the Earth around the Sun in all its history. However, we can obtain a simpler discrete equation in the following way. Let us define

$$y_t = y(2\pi t^+)\ ,\quad p_t = \frac{dy}{d\phi}\,(2\pi t^+)\, ,$$

$$C = \cos 2\pi Q,\qquad C = \sin 2\pi Q\ .$$

Equation (1.3) can then be written as

$$y_{t+1} = C\ y_t + (S/Q)\ p_t\ ,$$

$$p_{t+1} = -SQy_t + C\ p_t + BF(y_{t+1})\ ,$$

and after eliminating p one has

$$y_{t+1} + y_{t-1} = 2C\ y_t + (BS/Q)\ F(y_t)\ , \qquad (1.4)$$

an algebraic equation easy to solve. In Figure 1.3 we can see a typical diagram (y_{t+1}, y_t) for $F(y) = 2(1-\exp(-y^2/2))/y$, although many functions give rise to similar properties. The lower part is an enlargement around one of the hyperbolic points. All the chaotic points which "fill" a bidimensional region belong to the same orbit.

(d) The Rayleigh-Bénard Convection

This interesting phenomenon ((Bergé et al., 1984),(Schuster, 1984)) was discovered by Bénard in 1900 and explained by Lord Rayleigh in 1916. If a fluid layer is placed between two horizontal plates such that the lower one is hotter, the difference of the temperature being, there is a competition between

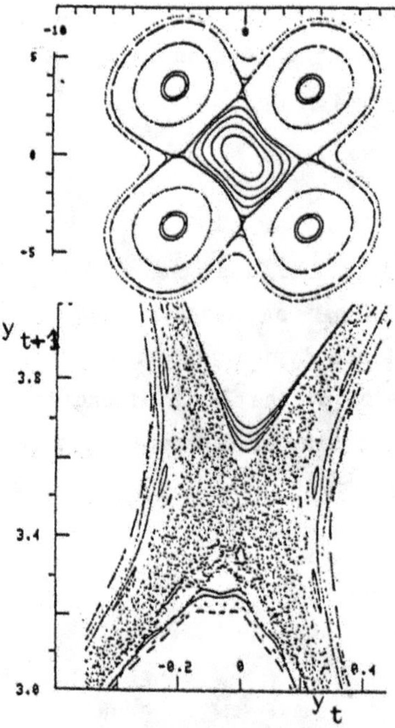

Figure 1.3. Phase space of Eq.1.4 (after Eminhizer, 1980[11]).

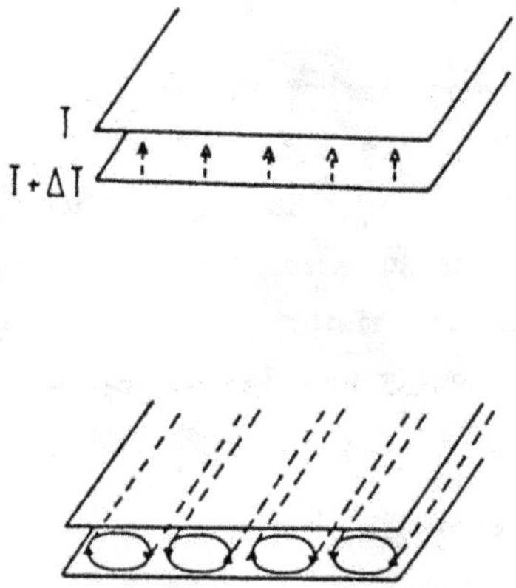

Figure 1.4. The Rayleigh-Bénard convection.

the tendency of the hotter fluid to rise and that of the cold one to fall (Figure 1.4). The viscosity is opposed to these tendencies and is able to impede any motion up to a certain threshold of ΔT. But, above this critical value, the equilibrium becomes unstable and the fluid begins to move with the classic convection rolls. The characteristic time for the equalization of the temperature by thermal diffusion is $\tau_T = d^2/D_T$, where d is the distance between the plates and D_T the thermal diffusivity. On the other hand, the characteristic time of the equalization through motion is $\tau_m = \eta/(\rho_0 g \alpha d \Delta T)$, where η is the viscosity, ρ_0 the density, g the acceleration of the gravity and α the dilatation coefficient. In order that a permanent effect be produced, the ratio of these quantities, called the Rayleigh number, must be greater than a certain critical value R_0, that is,

$$Ra = \tau_T/\tau_m = \frac{\rho_0 g \alpha d^3}{\eta D_T} \Delta T > Ro. \qquad (1.5)$$

The Rayleigh number is the control parameter. In order to modify it, the experimenter changes the difference ΔT. Above R_o the convection rolls appear and above another critical value the regularity disappears and the motion does not follow any recognizable pattern.

(e) The Belousov-Zhabotinsky Reaction

In 1958, the Russian chemist Belousov observed an oscillating chemical reaction, made apparent by an alternation of the color of the solution, which changed between yellow and colorless (In 1921 Bray, in California, had made similar observations, but nobody paid attention to his work). Some years later, Zhabotinsky studied the phenomenon in detail in his doctoral thesis. In the so called Belousov-Zhabotinsky reaction, an organic molecule (malonic acid) is oxidized by bromate ions, the process being catalyzed by a redox system (Ce^{4+}/Ce^{3+})[12,13] (see also {Schuster, 1984}). The basic reactants are

$$Ce_2(SO_4)_3,\ NaBrO_3,\ CH_2(COOH)_2,\ H_2SO_4,$$

to which a color indicator is added. The reaction is very complex, with as many as 18 elementary steps, and depends on several control parameters such as the temperature and the mean residence time in the open reactor. The variables of the associated dynamical system are the concentrations of the reactants. It is usual to observe that of the ion Ce_4^+, this being specially easy because of its strong absorption of light of 340nm. For some values of the control parameter, the concentration of Ce_4^+ (and that of the other chemicals) oscillates periodically. But above a certain critical value the behaviour is chaotic.

(f) The Bernoulli Shift

The Bernoulli shift[14] (see also {Arnold and Avez}, 1978) is a discrete dynamical system in [0,1), defined as

$$x_{n+1} = \sigma(x_n) = 2\, x_n \pmod 1 . \qquad (1.6)$$

It is convenient to use the binary representation of x:

$$x_o = 0.a_1a_2a_3 \ldots a_n \ldots = \sum_1^{\infty} a_k 2^{-k},$$

since it is then clear that the action of σ consists in erasing the first digit

$$\sigma(0.a_1a_2\ldots) = 0.a_2a_3a_4\ldots .$$

That of σ^k is therefore the elimination of the first k digits. This strongly chaotic system has the following properties:

(i) It has sensitive dependence on the initial conditions, because the errors double at each step. If the distance between two "seeds" x_o and x_o' is smaller than $1/2^n$, they have the first n digits in common, but they differ in the (n+1). Consequently the sequences generated by them will be completely different after the iteration number (n+1). If we only know the first n digits of x_o we know nothing about x_{n+1}.

(ii) Let us consider the coarse-grained evolution based in the partition of the phase space into the two intervals $L\equiv[0,\frac{1}{2})$, $R\equiv[\frac{1}{2},1)$. To every trajectory x_n we associate a sequence of L's and R's, depending on where x_n is. It is certainly isomorphic to the binary representation of x_o. For instance to 0.1001011... there corresponds RLLRLRR... If we identify R with heads and L with tails, x_o is equivalent to tossing a coin an infinite number of times. In fact, as none of the two sequences, digits of x_o or heads or tails, follow in general a regular pattern, if we are given a sequence of R's and L's, we cannot tell how it has been obtained.

(iii) The system is ergodic. This means that the trajectory generated by any seed, with the exception of a set of measure zero (the rationals), passes arbitrarily close to any number x in [0,1). To be specific, if we require that it passes closer than $\epsilon=2^{-n}$, it is necessary that there is an integer m, such that the first n digits of x_m coincides with those of x. But this is precisely the case, since the decimal representation of any irrational contains any finite sequence of digits an infinite number of times.

(iv) In order to predict the sequence x_k for all k it is necessary to know the initial condition x_o with an infinite number of digits, that is with zero error.This is precisely the reason for the unpredictability of a chaotic system: the determinism of the equation of motion does not allow an effective prediction of the trajectory if the initial conditions are not known with infinite precision; but this is impossible in data obtained by measurement or calculation, as this would imply an infinite amount of information.

It is easy to understand the mechanism which is responsible for the chaos. It consists in the two effects which appear schematically represented in figure 1.6:

(a) Because of the product by 2, the intervals are **"stretched"**, the errors are amplified and the neighbouring points are separated.

(b) Because only the decimal part is taken, the stretched total interval is **"folded"** and the previously separated points are intermingled.

This combination of stretching and folding is similar to the shuffling of a deck of cards. In fact, in this case, the deck is first expanded, so that the cards are slightly separated, and is later folded on itself. For this reason this mechanism, which is present in general chaotic systems, could be called shuffling effect. Curiously enough, many natural deterministic systems behave as cards which are shuffled.

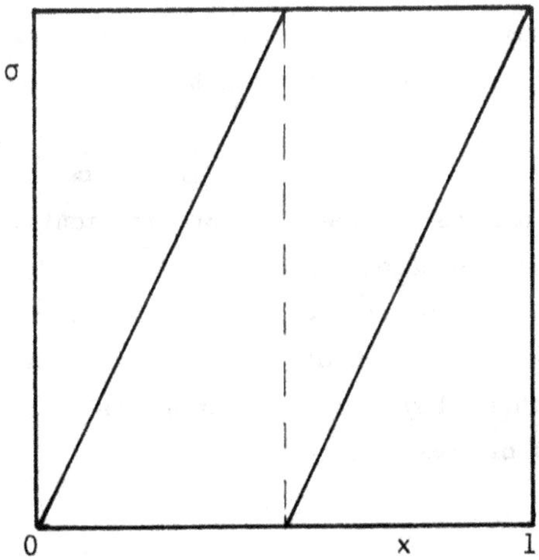

Figure 1.5. The Bernouilli shift $\sigma(x)$.

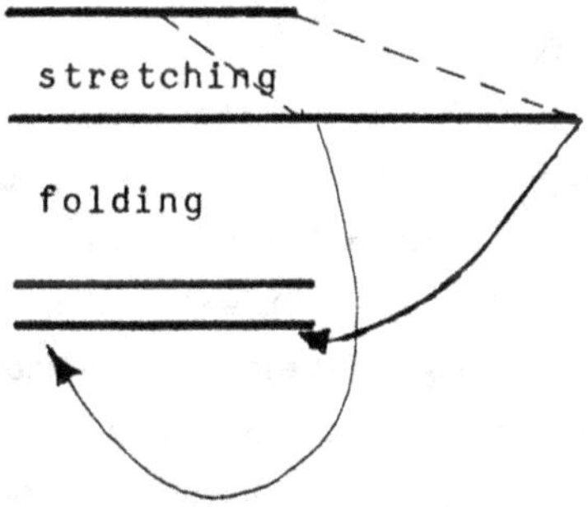

Figure 1.6. Stretching and folding in the Bernouilli shift.

1.3 SEVERAL QUESTIONS ABOUT CHAOS

All this suggests several questions:

(i) What is chaos? How to know, from its equation of motion, if a system is regular or chaotic?

(ii) How general a phenomenon it is? Or, in other words, is it the exception or the rule?

(iii) Is it just a mathematical problem or, on the contrary, does it affect the very concept of physical law?

(iv) As the physics of regular systems has achieved a great success in the description of nature, which is the relation between order and chaos?

In these notes we will try to answer these questions. Concerning the first one, let us say that no formal definition of chaos will be given, since none is completely adequate. Instead of that, we will follow a descriptive method and we will give some characteristics which are usually interpreted as its manifestation. An important problem is to determine if a system is regular or chaotic, just from the knowledge of the equation of motion. As it was stated in the Introduction, there are some criteria which can be applied to Hamiltonian systems, but which are useful only in some cases.

To answer the second question, it must be said that chaos is ubiquitous. Again, we can be more precise in the case of Hamiltonian systems, since Siegel proved two theorems (Siegel 1941, 1954[15]; Moser 1968, 1973[16,17]) which characterize the meaning of ubiquity, in these cases. Siegel considered Hamiltonians which are analytic functions of the coordinates and the momenta, so that they can be written as (for simplicity we will assume two degrees of freedom)

$$H(x,y,\dot{x},\dot{y}) = C_{ijkl}\, x^i y^j \dot{x}^k \dot{y}^l \quad , \tag{1.7}$$

and defined a neighborhood of H as the set of Hamiltonians H', such that their coefficients satisfy

$$|C'_{ijkl} - C_{ijkl}| < \varepsilon_{ijkl} \quad , \tag{1.8}$$

where the ε's are arbitrary positive numbers. In 1941, he showed that any neighborhood of H contains another H' , which is nonintegrable, which implies that nonintegrable (and therefore possibly chaotic) Hamiltonians are dense in the set of those which are analytic. In 1954, he was able to prove that integrable system are not dense in this sense. We may express this result by saying that integrability and order are the exception (in the sense of Baire category)rather than the norm in the case of Hamiltonian systems. Therefore, and this is very important, an approximation to an integrable system may be chaotic and, as the number of systems which can be exactly solved is very small, chaos is unavoidable in everyday work in physics.

In the case of nonconservative systems, there are no results analogous to those of Siegel, but it seems that, if they are nonlinear, they will be completely regular only excepcionally.

The third question is a very relevant one. It could be thought that chaos presents no more than a mathematical difficulty, which will be overcome in a few years by the development of suitable calculational techniques. Quite on the contrary, some authors believe that it presents a far reaching problem that affects the very concept of physical law. This is because one of its immediate consequences is that chaotic systems cannot be predicted for times greater than a value which depends on the initial data and on the intensity of the stochasticity. I. Prigogine speaks of a "time window" out of which it is impossible to make reliable predictions, since this would need an infinite amount of information, which can only be handled with superhuman abilities. This limitation has important consequences on our image of nature and has been compared to the one imposed by the Heisenberg relations in quantum theory.

1.4 THE SIGNS OF CHAOS

As it was stated in the preceding section, no attempt to give a formal definition of chaos will be made. But, following a descriptive method, we will examine some of its most characteristic properties.

1.4.1 The Aspect of the Signal

In some cases a signal, obtained either by measurements or by numerical analysis, looks chaotic, this being an indication of nonregular motion. However, this criterion is not always precise and reliable, since it is often difficult to distinguish between a regular but complicated pattern and fully irregular behaviour.

1.4.2 The Power Spectrum

It is well known that according to the theory of Fourier series the frequency spectrum of a periodic function consists of a fundamental and its harmonics. In some cases, the time evolution is more complex, but still regular, and there are, instead of one, two or several basic frequencies, together with all their linear combinations with integer coefficients. The frequency spectrum is therefore discrete. This is a characteristic of regular motion, while deterministic chaos has a continuous spectrum. It is true that, if the number of frequencies is large, it may be difficult to distinguish between the two cases, but this is just a practical problem. In the following we will identify the discrete spectrum with regularity and the continuous one with chaos. But we must examine in detail this question.

Even in the case of continuous systems described by differential equations, the numerical data are based on a discretization of time, while the measurents are usually made at regular intervals of time. We will, therefore, concentrate on sequencies such as

$$x_j, \quad j = 1,2,\ldots,n,$$

which are measures or theoretical approximations to $x(j\Delta t)$, x and Δt being any variable and an increment of time. If there are n values of x, the time interval is $(0,t_{max}) \equiv (0,n\Delta t)$. Although we will not consider the mathematical details, it is understood that $x(t)$ is integrable and square-integrable in the interval.

Discrete Fourier transform

Given the sequence $x_1,\dots,x_n$, its Fourier transform $\hat{x}_1,\dots,\hat{x}_n$ is defined by

$$\hat{x}_k = \frac{1}{\sqrt{n}} \sum_{i=1}^{n} x_j \exp\{-i\frac{2\pi jk}{n}\} \tag{1.9a}$$

(to simplify,we have chosen the units such that $\Delta t=1$; otherwise we should write $n\Delta t$ and $j\Delta t$ instead of n and j in (1.9a)). The inverse transformation is

$$x_j = \frac{1}{\sqrt{n}} \sum_{k=1}^{n} \hat{x}_k \exp\{ i\frac{2\pi kj}{n}\} . \tag{1.9b}$$

(i)The two sequences are related through the Parseval-Plancherel theorem, which states that

$$\sum_j |x_j|^2 = \sum_k |\hat{x}_k|^2 . \tag{1.10}$$

(ii) While the argument of x is a time, and $x_j = x(j\Delta t)$, that of $\hat{x}$ is a frequency and $\hat{x}_k = x(k\Delta f)$. The frequency increments are $\Delta f = 1/t_{max} = 1/n\Delta t$, the maximum frequency being $1/\Delta t$.

(iii) If the variable x is real,

$$\hat{x}_k = \hat{x}^*_{n-k} . \tag{1.11}$$

(iv) The sequence $\{x_j\}$ can be considered as a part of an infinite periodic one with period n, so that $x_j = x_{j+n}$. At times, it is convenient to take the limit $n\to\infty$.

Frequency spectrum

The intensity of the term with frequency k is characterized by the important function

$$E_k = |\hat{x}_k|^2 .$$

Its dimensions depend on those of x, but, in spite of that, it is usually called power spectrum. This is because, in the important case of the reception of electromagnetic waves in an antenna with linear detector, E_k is directly related to the transported power.

Different types of Fourier spectra

(i) Periodic signal

Let $x(t)=x(t+T)=x(t+2\pi/\omega)$. Let us also take $t_{max}=pT$ or, equivalently, $\Delta f=1/pT$. It is easy to show that if x(t) is sinusoidal,

$$\hat{x}_p = \hat{x}^*_{n-p} \neq 0 , \quad \hat{x}_k = 0 \text{ if } p \neq k \neq n-p , \tag{1.12}$$

that is, there is only one component (plus the one due to relation (1.11)). If x(t) is a nonsinusoidal periodic function there are also components at the frequencies 1/T, 2/T,.. But, what if t_{max}/T is not an integer, as is the case most of the time? What happens then is that frequencies widen, so that each line correspond to several neighbouring values of k (see {Bergé et al., 1978}). In any case, the spectrum of a periodic signal consists of a line at frequency 1/T and lines at its harmonics m/T.

(ii) Quasi-periodic signal

A quasi-periodic function of time is one which can be written as

$$f(\omega_1 t, \ ,\omega_k t, \ldots ,\omega_r t) , \tag{1.13}$$

where f is periodic and has period 2π in all the r arguments $t_k=\omega_k t$; it is also called r-periodic. The frequencies are $f_k=\omega_k/2\pi$ and the periods $T_k=2\pi/\omega_k$. The trajectory in phase space corresponding to a quasi-periodic function is situated in a torus T^n. For instance, in the biperiodic case, (x,y,z) is in a torus if $x=x(\omega_1 t,\omega_2 t)$, $y=\dot{x}$, $z=\ddot{x}$.

The Fourier spectrum of a quasi-periodic function may be very complex, since it includes all frequencies

$$m_1f_1+m_2f_2+\ .\ .\ +m_rf_r\ ,$$

where the m_i are integers. Let us consider the biperiodic case with continuous t and f; every component of the spectrum is a peak with $f=|m_1f_1+m_2f_2|$ which is usually referred to as (m_1,m_2). Let $M=|m_1|+|m_2|$.There are essentially two different alternatives:

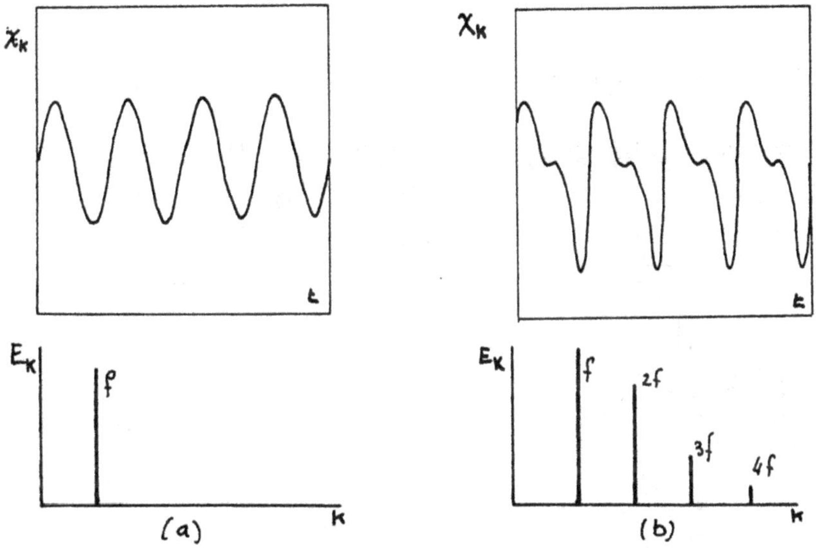

Figure 1.7. Signal and power spectrum in the cases sinusoidal (a) and periodic (b).

(a) f_1/f_2 is an **irrational** number.

In this case, the numbers $|m_1f_1+m_2f_2|$ form a set which is dense in R^+, so that there is a frequency arbitrarily close to any positive number. However, the situation is generally simpler and the amplitudes decrease fast for large values of M, so that in most cases it is enough to consider only those of low M. As an example, let us take the function $y=\sin(\omega_1t)\sin(\omega_2t)$. It can be written as

$$y(t)=\frac{1}{2}\cos\ ((\omega_1-\omega_2)t)\ -\frac{1}{2}\cos\ ((\omega_1+\omega_1)t)\ ,$$

so that it has only the peaks (1,-1) and (1,1), both with M=2.

(b) f_1/f_2 is a **rational** number equal to n_1/n_2 with n_1 and n_2 integers.

In this case, the biperiodic function is simply periodic with period $T = n_1 T_1 = n_2 T_2$, so that all the lines in the spectrum are harmonics of the fundamental frequency $f_0 = f_1/n_1 = f_2/n_2$.

In general, one of the following situations hold:

(a) The r frequencies are incommensurable because there are no integer numbers such that

$$n_1 f_1 + n_2 f_2 + \ldots + n_r f_r = 0 . \qquad (1.14)$$

It is then said that the motion is truly or strictly quasi-periodic.

(b) There are integers n_i, such that (1.14) is satisfied. The motion is then simply periodic. It is then said that there is frequency locking.

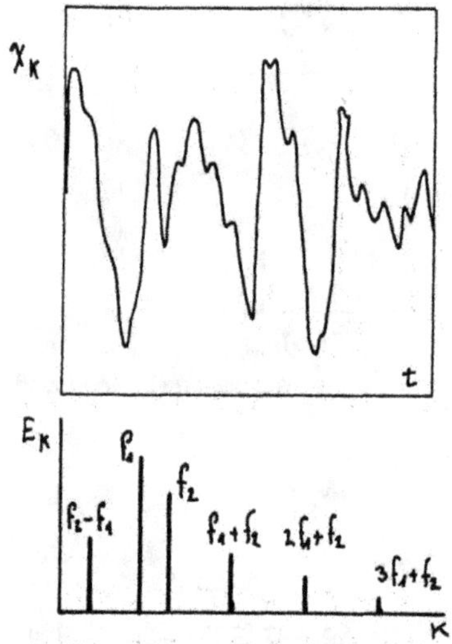

Figure 1.8. Quasi-periodic signal and its power spectrum.

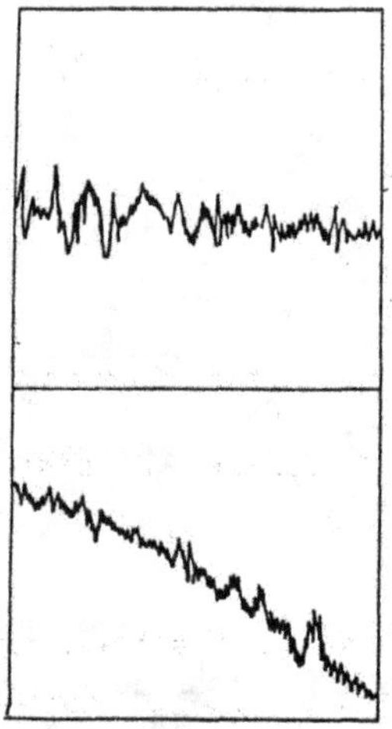

Figure 1.9. Aperiodic signal and its power spectrum.

(iii) Aperiodic signal

When a function x(t) is not periodic it is called aperiodic or non-periodic. Its frequency spectrum has a continous portion without strong maxima. It is then said that there is broadband noise. But it must be stressed that in many occasions it is difficult to distinguish between a discrete spectrum with many lines and a really continuous one.

Aperiodic signals in systems with few degrees of freedom correspond to deterministic chaos. It is convenient, because of methodological reasons, to distinguish this from the case in which there are many degrees of freedom, to which the deterministic techniques used in classical mechanics (e.g., the numerical integration of the equations) cannot be applied practically. Another comment is in order: when the randomness is maximum, the function E_k is constant. The signal is then called "white noise", since all the frequencies have the same amplitude, as occurs for white light. In deterministic chaos, the motion is less random than in white noise, and this is expressed by E_k depending on k.

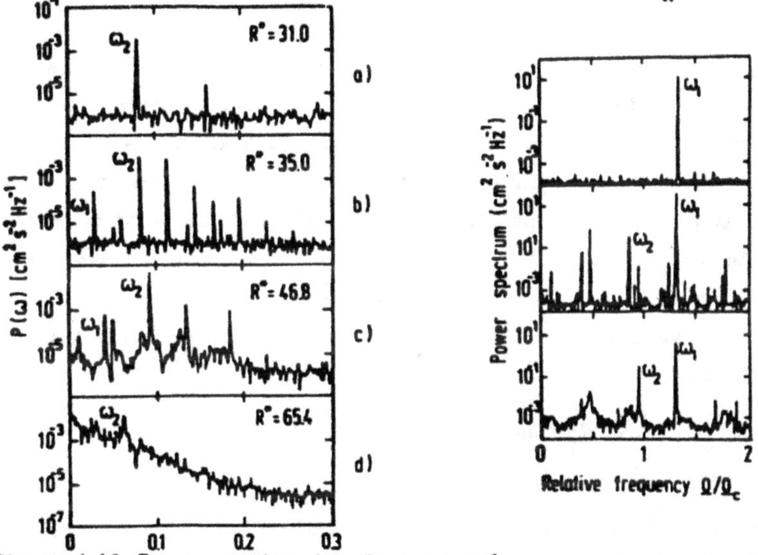

Figure 1.10. Power spectra in Rayleigh-Bénard convection (left) and in Couette flow (right) (both after Swinney and Gollub, 1978 [18]).

As an illustration, we can see in Figure 1.10 the power spectra of the fluid velocity in a Rayleigh-Bénard experiment, measured by means of the Doppler effect in light, and of a Couette flow. In the upper parts, the signals are periodic and the spectra consist of a fundamental frequency and its harmonics. In the middle ones, there are two incommensurable fundamental frequencies and the motion is quasi-periodic. In the lower parts, the behaviour is chaotic, as it is clear from the continuous aspect of E_k. At the end of chapter 2 we will consider again these figures.

1.4.3 The Correlation Function

The correlation function of the signal x_j is defined as

$$C_m = \frac{1}{n}\sum_j x'_j\, x'_{j+m} \,, \tag{1.15}$$

$$x'_j = x_j - \bar{x} = x_j - \frac{1}{n}\sum_k x_k \,.$$

Frequently when the number of data is very big, the limit $n\to\infty$ is taken. It is clear that

$$C_m = \frac{1}{n}\sum_j x_j\, x_{j+m} - \bar{x}^2 = \psi_m - \bar{x}^2 \,. \tag{1.16}$$

C and ψ are discretizations of functions of time $C_m = C(m\Delta t)$, $\psi_m = (m\Delta t)$.

The correlation function expresses how much x_j differs from its mean value. More precisely, it is a measure of how much the difference $x_j - \bar{x}$ resembles its own value m intervals Δt before. If C_m is large, the signal changes little after m steps, being thus still predictable after a time $m\Delta t$. On the other hand, if it is small or zero, x_{j+m} is not close to x_j on the average and prediction is not possible.

Let us calculate ψ_m (which is the same as C_m if the mean value of the signal vanishes):

$$\psi_m = \frac{1}{n^2}\sum_j \sum_{k,k'} \hat{x}_k\, \hat{x}_{k'} \exp\{i\frac{2\pi}{n}(jk+(j+m)k'\}$$

After summing over j and k', one obtains (remember (1.11)):

$$\psi_m = \frac{1}{n}\sum_k |\hat{x}_k|^2 \exp(i\frac{2\pi mk}{n}) = \frac{1}{n}\sum_k E_k \cos\frac{2\pi mk}{n}$$

and analogously

$$E_k = |\hat{x}_k|^2 = \sum_m \psi_m \cos\frac{2\pi mk}{n} \,,$$

where a term $\frac{1}{n}\hat{x}_n^2$ has been neglected, as its limit when $n\to\infty$ is zero.

As we see, ψ_m and E_k are Fourier transforms of one another (more precisely, according to our normalization $\sqrt{n}\,\psi_m$ and E_k. This result is known as the Wiener or Wiener-Khinchin theorem. It indicates that there are some similarities with the situation in quantum mechanics: if ψ_m decreases fast, it is very localized and E_k must be a wide function, in other words, there must be broadband noise. Conversely, in the extreme situation in which E_k has only one frequency, ψ_m can not go to zero and must be periodic. Let us consider some examples.

Unidimensional harmonic oscillator

The signal is $x(t) = A \cos \omega t$, from which the correlation function is (an integral is taken instead of a sum)

$$C(\tau) = \lim_{T\to\infty} \frac{1}{T} \int_0^T x(t)x(t+\tau)dt = \frac{A^2}{2} \cos \omega\tau ,$$

so that $C(\tau)$ does not decrease as a consequence of the regularity of the motion.

Bernouilli shift

To calculate the correlation function in this case, we will use the concept of invariant density. This is a function $\rho(x)$ such that, if we place at $t=0$ a large number of systems in phase space with density proportional to $\rho(x)$ (between x and $x+\Delta x$ there are $N\rho(x)$ systems, where N is a normalization constant), this density does not change when all these systems follow their trajectories. In principle, there can be several such functions. The case in which there is only one, up to the product by constants, is specially interesting, since the system is then ergodic.

Let us take an initial value x_0. We define $\rho(x)$ as

$$\rho(x) = \lim_{n\to\infty} \frac{1}{n} \sum_i \delta(x-f^i(x_0)). \tag{1.17}$$

If this limit exists and is independent of x_0, then the time averages can be written as averages in phase space with respect to $\rho(x)$:

$$\lim_{n\to\infty} \frac{1}{n} \sum_i g(x_i) = \int_0^1 dx\ \rho(x)g(x).$$

In order to calculate $\rho(x)$, let us take an initial density ρ_0, which successively changes to $\rho_1, \rho_2, \ldots, \rho_n$, so that

$$\rho_{n+1}(y) = \int_0^1 dx\ \delta(y-f(x))\rho_n(x) . \tag{1.18}$$

As we see, (1.17) is invariant under (1.18), from which the invariant density satisfies the integral equation of Frobenius type

$$\rho(y) = \int_0^1 dx\ \delta(y-f(x))\ \rho(x)\ . \tag{1.19}$$

In the case of the Bernouilli shift,

$$f(x)=\sigma(x)= \begin{cases} 2x & \text{if} \quad 0 \leq x < 1/2\,, \\ 2x-1 & \text{if} \quad 1/2 \leq x < 1\,, \end{cases}$$

and (1.19) takes the form

$$\rho(y)= [\rho(y/2)+\rho((y+1)/2)]/2\,,$$

the normalized solution of which is $\rho=1$. That it is the only one can be easily proved by inserting any normalized ρ_0 in (1.18) and checking that the limit when $t\to\infty$ is always 1.

The correlation function can be written, in terms of ρ, as

$$C_m = \int dx\ \rho(x)\ x\ f^m(x) - \{\int dx\ \rho(x)\ x\}^2 ..$$

In this case,

$$C_m = \int dx\ x\sigma^m(x) - \{\int dx\ x\}^2 .$$

The second integral is 1/2. To calculate the first one, let us note that the graph of the function σ^m consists of 2^m segments with slope 2^m, in each one of the subintervals $[a_{k-1}, a_k)$, $a_k = k/2^m$, $k=1,2,..2^m$ (see Figure 1.11). We thus have

$$I= \int dx\ x\ \sigma^m(x) = \sum I_k\ ,$$

$$I_k = \int_{a_{k-1}}^{a_k} x\ \sigma^m(x)dx = \int_0^{2^{-m}} (a_{k-1}+z)2^m z\ dz$$

$$= 2^m\ (a_{k-1}\ z^2/2 + z^3/3) = 2^{-2m}(k/2-1/6)$$

and hence $I = 2^{-m}/12 + 1/4$, and the correlation function is

$$C^m = 2^{-m}/12 ,$$

which decreases exponentially. It should be stressed that there are systems in which the correlation functions decrease faster and are even zero for m>0. As an example, we can mention the triangular map (Figure 1.12):

$$\Delta(x) = 1 - 2|1/2 - x| .$$

This system is also ergodic and its invariant density is again $\rho=1$. By the same procedure as before, it is easy to prove that $C_m = \delta_{m,o}/12$.

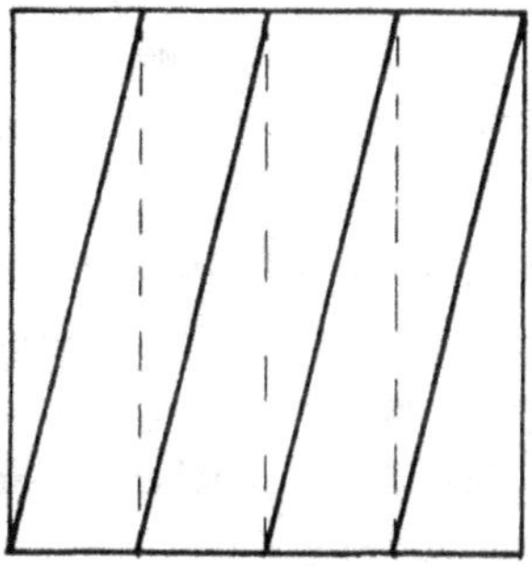

Figure 1.11. Shape of σ^2.

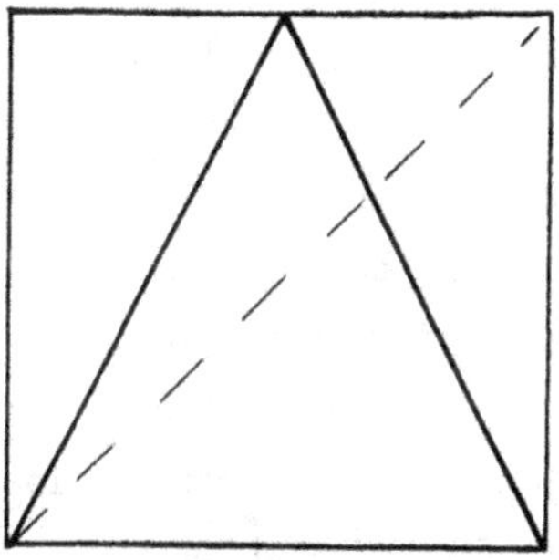

Figure 1.12. The triangular map.

1.4.4 The Poincaré Map

Poincaré invented a mathematical technique which is very useful in the investigation of the behaviour of system, as it allows one to reduce the problem to a simpler one with fewer dimensions. For the sake of argument, let us consider a conservative system with two degrees of freedom. The equations can then be written as

$$\dot{x}_i = F_i(x), \quad i=1,2,3,4 \quad ; \quad (x_1,x_2,x_3,x_4) \equiv (x,y,p_x,p_y) .$$

The trajectories are contained in the energy shell H(x,p)=E, so that the accessible space is a three-dimensional manifold, which can be represented as in Figure 1.13, with coordinates x_1, x_2 and x_3, the fourth one, x_4, being determined , frequently up to its sign, by the condition H=E. Let us now consider the plane S, defined as y=0, with coordinates x, p_x. A generic trajectory with $\dot{y}>0$ cuts this plane at the points $Z_o, Z_1, \ldots, Z_n, \ldots$, so that any of them is determined by the preceding one. The plane is called Poincaré section or surface of section and the Poincaré map is defined as the corresponding mapping of the plane into itself:

$$Z_k \to Z_{k+1} = P(Z_k) \ , \qquad (1.20)$$

so that $P^n(Z_n)=Z_n$. In the general case, the Poincaré map is more complex, since if the system is nonconservative and the dimension of phase space is n, a hypersurface of n-1 dimensions must be used as the section. But we can still obtain interesting conclusions by studying the projection of the trajectories on a suitably chosen three-dimensional subspace. For this reason, bidimensional Poincaré sections are always employed in practice.

Another necessary remark is that the Poincaré map retains many of the properties of the original flow. If it is conservative, the same is true for the Poincaré map, the area being conserved. If it is dissipative, there is both contraction of volumes in phase space and of areas in the surface of section.

Let us now see how from the study of the Poincaré map some important properties of the trajectories can be deduced (Figures 1.13, 1.14 and 1.15). For instance, if a solution is periodic, the curve in phase space is closed and the number of cuts is finite, P being also a periodic map. In the simplest case, there is only one cut which is a solution of P(Z)=Z and the stability of the solution is the same as that of this fixed point. To first order, the Poincaré map is given by its linearization

$$P(Z+\delta) = P(Z) + M\delta,$$

where M is the Floquet matrix, that is ,

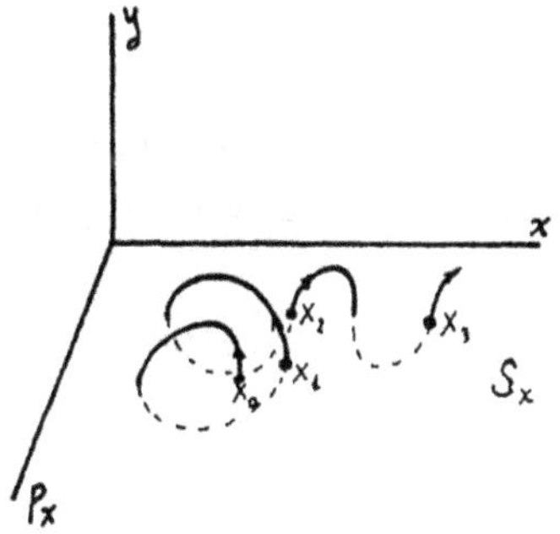

Figure 1.13. A Poincaré map.

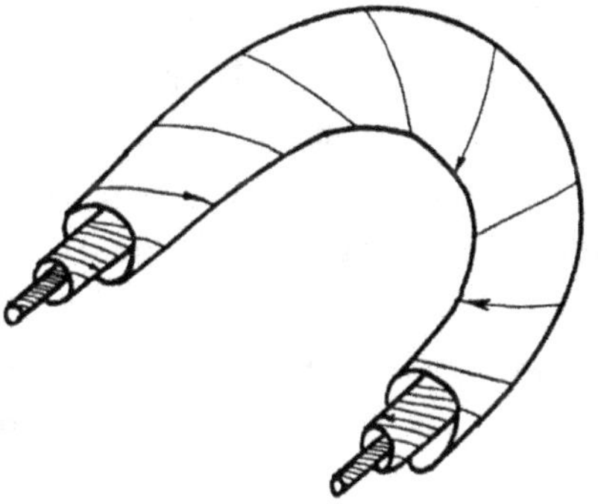

Figure 1.14. Trajectories in tori.

$$M = \left(\frac{\partial P}{\partial x}\right)_Z .$$

The linear stability of a periodic trajectory is thus given by the eigenvalues of M. If all of their moduli are smaller than one, it is an attractor limit cycle; otherwise it is an unstable solution.

Let us now take the case of a biperiodic solution with frequencies f_1 and f_2. As we have noted, the trajectory is contained in a torus T^2, which can be considered as a submanifold of R^3, and turns around each one of its two "axes". The cuts are produced at regular time intervals equal to $T_1 = 1/f_1$ and are placed in the intersection of the torus and the surface of section, a closed curve which is invariant under P. The aspect of the section depends specially on the relation f_1/f_2. If it is irrational, the trajectory covers ergodically the surface of the torus, so that the cut points are densely placed along the invariant curve C. If $f_1/f_2 = n_1/n_2$, n_1 being integers, the solution is periodic and closes upon itself after executing n_1 and n_2 turns around the two "axis" of the torus. The Poincaré section of this solution consists then of n_1 points $Z_1, \ldots, Z_n$ and

$$Z_i = P(Z_{i-1}) .$$

If the solution is aperiodic, the cut points appear in the surface in a disordered manner, with no recognizable pattern, and are not in any curve, but, on the contrary, explore a two-dimensional region, a property referred to as "space filling".

To summarize, the Poincaré map allows the reduction of any dynamical system to a discrete map with fewer dimensions. Moreover ,it exhibits the main characteristics of the motion in a very expressive manner.

As an illustration, let us consider the system of Hénon and Heiles (1964),[19] in which a point particle moves in two dimensions in the potential

$$V(x,y) = (x^2+y^2)/2 + x^2y - y^3/3,$$

which in polar coordinates has the form

$$V(r,\phi)= r^2/2 + \frac{1}{3} r^3 \sin 3\phi$$

and which is used in the study of triatomic molecules and in stellar dynamics. In Figure 1.16, the corresponding Poincaré maps in the y,p_x - plane for four energy values are shown. The result of a perturbative calculation[20], at left, is compared with the numerical analysis of the equation of motion, at right. If E=1/24 or E=1/12, the perturbative results agree well with the numerical ones, and inthis approximation the map consists of invariant curves which are the intersections of the tori and the plane of section. But Hénon and Heiles found that if the energy is greater than 1/9, the situation becomes completely different. The perturbative calculations are no longer good and only some of the tori remain. In most cases the successive iterations cover in an erratic and irregular way the surface of section. For instance, all the points which are not clearly in an invariant curve in the cases E=1/8 and E=1/6 belong to just one trajectory.

It is worth mentioning that the Hénon-Heiles system is the approximation, up to cubic terms in the Hamiltonian, to an integrable system with regular behaviour, the so called Toda chain which has the potential

$$V(x,y) = (\exp(2y+2\sqrt{3}\,x) + \exp(2y-2\sqrt{3}\,x) + \exp(-4y))/24 - 1/8$$

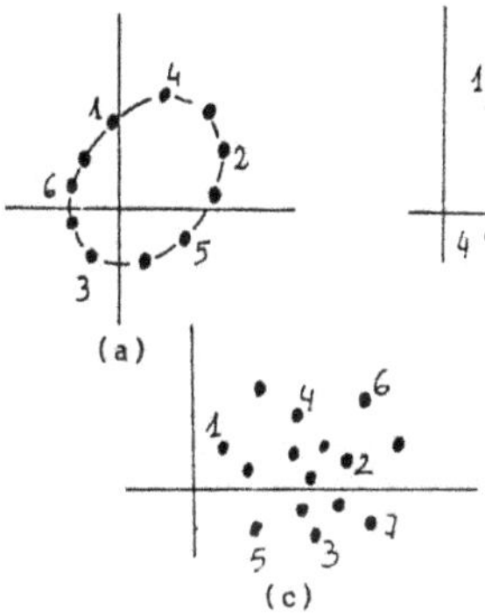

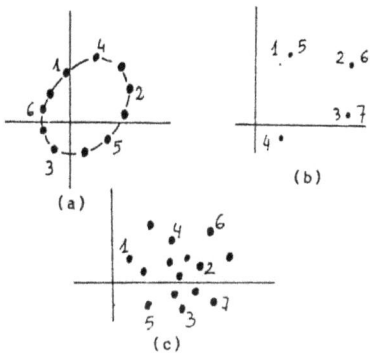

Figure 1.15. Poincaré sections for quasi-periodic(a),periodic (b) and aperiodic motions(c).

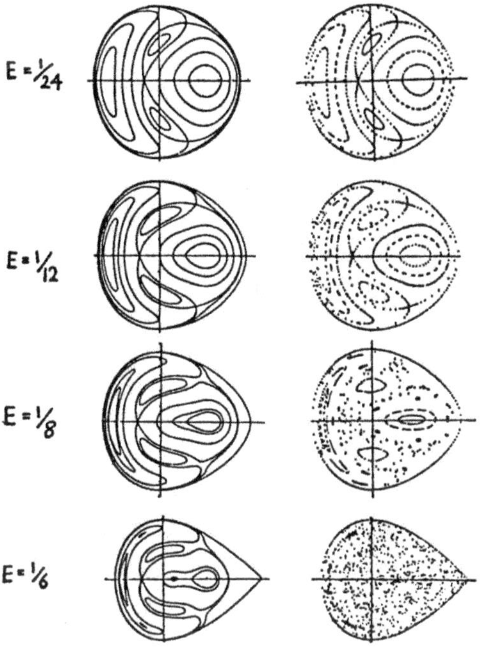

Figure 1.16. Poincaré sections of the Hénon-Heiles system,see text (after Gustavson,1966 [20]).

and gives a model of a simple solid with only three particles in a ring. This should not be surprising, since the above mentioned two theorems by Siegel warn that an approximation to a regular system may introduce chaos.

1.4.5 The Liapunov Exponents

Let us consider the discrete system

$$x_{n+1} = f(x_n) = a\, x_n, \quad |a| < 1.$$

As an effect of the iterated application of $f(x)$, the points which are close separate. More precisely, if $|x_o-y_o|=\epsilon$, then $|x_n-y_n|=|a|^n|x_o-y_o|$. In the case of a general function f, we may write

$$|f^n(x_o+\varepsilon) - f^n(x_o)| = \varepsilon \exp(n\,\lambda(x_o)),$$

which, in the limit $\epsilon\to 0, n\to\infty$ gives

$$\lambda(x_o) = \lim_{n\to\infty}\lim_{\varepsilon\to 0} \frac{1}{n}\log\left|\frac{f^n(x_o+\varepsilon) - f^n(x_o)}{\varepsilon}\right|, \tag{1.21}$$

that is ,

$$\lambda(x_o) = \lim_{n\to\infty} \frac{1}{n}\log\left|\frac{df^n(x_o)}{dx_o}\right|. \tag{1.22}$$

$\lambda(x_o)$ is called a Liapunov exponent and gives a measure of the exponential separation of near trajectories: they move away by a factor $\exp(\lambda)$ at each step. Because of the chain rule,

$$\left.\frac{df^2(x)}{dx}\right|_{x_o} = f'(x_1)\,f'(x_o),$$

and, in general

$$\left.\frac{df^n(x)}{dx}\right|_{x_o} = f'(x_{n-1})\,f'(x_{n-2}).f'(x_1)\,f'(x_o)\ ,$$

from which

$$\begin{aligned}\lambda(x_0) &= \lim_{n\to\infty}\frac{1}{n}\log\left|\prod_0^{n-1} f'(x_i)\right| \\ &= \lim_{n\to\infty}\frac{1}{n}\sum_0^{n-1}\log|f'(x_i)|\ .\end{aligned} \tag{1.23}$$

In the case of the linear function $f=ax$, $\lambda=\log|a|$. It is clear that a positive Liapunov exponent is a necessary condition for chaos. But, as the linear example shows, this is not sufficient; in addition, a folding nonlinearity is required, in order to have stretching and folding.

The consideration of this concept from the point of view of Shannon information theory [21] is very enlightening. Let us divide the unit interval into n equal parts and assume that a point can be in any of them with the same probability. According to Shannon, if we determine in which of these intervals it is, we gain an amount of information equal to

$$I_0 = -\sum \frac{1}{n} \log_2 \frac{1}{n} = \log_2 n.$$

This is a particular case of the definition of the information gain in determining in which state a system is if there are n possible states and the a priori probabilities of the system being in state i is p_i. Shannon proposed that this is

$$I = -\sum_1^n p_i \log_2 p_i .$$

A system with two equiprobable states ($p_i=½$) contains the minimal amount of information, which is equal to 1 and is called a "bit". Let us consider again the linear function $f(x)=ax$. The lengths and the errors change by a factor $a=|f'(x)|$ and this produces an information change

$$\Delta I = -\sum_1^{n/a} \frac{a}{n}\log_2 \frac{a}{n} + \sum_1^n \frac{1}{n} \log_2 \frac{1}{n} = -\log_2 a = \log_2 |f'(0)|.$$

If $|f'(x)|>1$, there is an effective loss of information because of the lesser resolution. In the case of a general map, the average loss of information per step is

$$\overline{\Delta I} = -\lim_{n\to\infty} \frac{1}{n} \sum_0^{n-1} \log_2 |f'(x_i)| . \qquad (1.24)$$

We see that it is proportional to the Liapunov exponent

$$\lambda(x_o) = -\log 2\ \overline{\Delta I} ,$$

so that if $\lambda>0$ there is information loss. This is because, as the prediction of the state of the system is necessarily less accurate, the information gained in its determination is smaller.

If the system has d dimensions, there are d Liapunov exponents, defined as

$$(\exp \lambda_1, \exp \lambda_2, \ldots, \exp \lambda_d) = \lim_{n\to\infty} [\text{eigenvalues of } \prod_0^{n-1} J(x)]^{1/n} ,$$

where

$$J(x) = \det \left(\frac{\partial f_i}{\partial x_j} \right) .$$

is the Jacobian of the transformation. To these eigenvalues correspond d eigendirections, along which distances expand by the factors exp λ_k. It is clear that the dominant effect is due to the greatest exponent which is thus the simplest to compute. In order to do that, the following procedure is normally used. Take a fiducial trajectory with initial value x_0 and another one which at t=0 is at a distance $\epsilon \ll 1$. Both are integrated for a while and the increase in their distance is calculated. Then the second one is chosen near the fiducial one and the process is repeated several times. Then the exponent is calculated by using (1.21) (Figure 1.17). In general, it is more difficult to calculate the other exponents.

Let us take two examples. First, the Bernouilli shift. As all the intervals double each step, it is clear that $\lambda=\log 2$, which is positive, as it must be in the case of a chaotic system. The second one refers to the oscillator with equations

$$\begin{aligned} \ddot{x}_1 + \omega^2 x_1 + \varepsilon x_1 x_2^2 &= 0 , \\ \ddot{x}_2 + x_2 + \varepsilon x_2 x_1^2 &= 0 . \end{aligned} \tag{1.25}$$

This system has both regular and chaotic orbits in a proportion which depends on the value of ϵ. In the Figure 1.18, the logarithm of the distance between a trajectory and a nearby one is represented versus time, both in the regular and in the chaotic regime. The saturation which is seen for large times is due to the distance becoming of the same order of magnitude as the accessible region of phase space.

1.4.6 The Kolmogorov Entropy

Let us recall the definition of entropy used in statistical mechanics:

$$S = -k \sum p_i \log p_i \, ,$$

where p_i is the probability of the system being in the state i. As it is well known, S is a measure of the disorder of the system and, from the point of view of Shannon information theory, it indicates the amount of information which is needed to determine the state of the system; S is thus a gauge of our ignorance of this state.

Kolmogorov used the same idea to define an indicator of the intensity of chaos, now called Kolmogorov entropy or K entropy, which is equal to the mean loss of information on the state of a system, when it

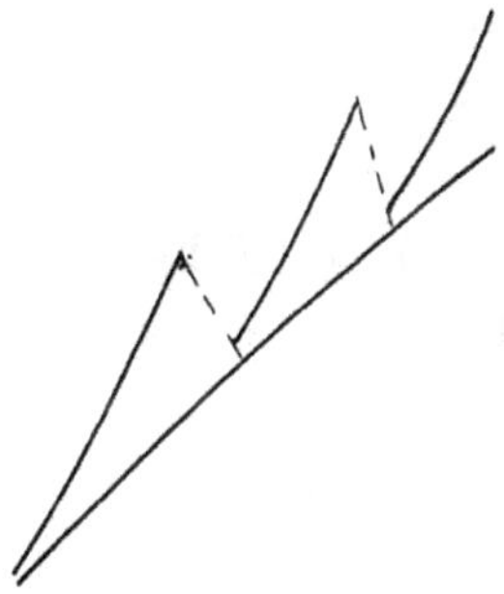

Figure 1.17. Calculation of the Liapunov exponent.

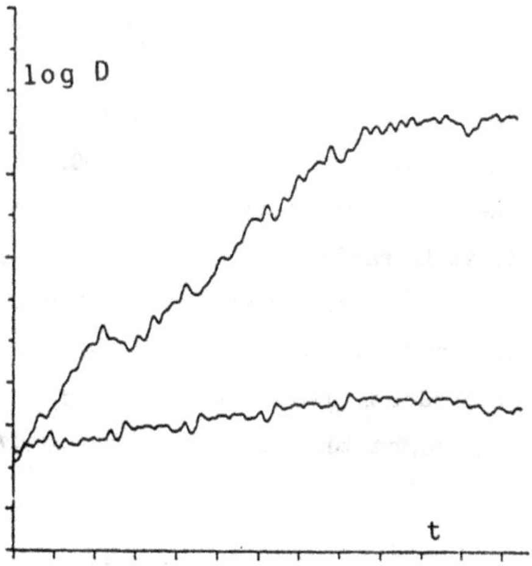

Figure 1.18. Separation in time of trajectories of the oscillator 1.25. A chaotic and a regular case are shown (after López, 1985 [22]).

evolves in time. Let us take a trajectory $x(t)$. If we make a partition of the phase space into cells of volume ϵ^d and measure or calculate its state at regular time intervals τ, we can associate to any trajectory a sequence $i_o, i_1, \ldots, i_n$, which indicates that at time $t=k\tau$ the system was at cell i_k. Let $P_{i_0 i_1 \ldots i_n}$ be the probability of this being so. Following Shannon, the quantity

$$K_n = -\sum_{i_o,\ldots,i_n} P_{i_o i_1 i_2 \ldots i_n} \log P_{i_o i_1 \ldots i_n}$$

is proportional to the information needed to determine the coarse-grained trajectory if only the probabilities are known. $K_{n+1}-K_n$ is therefore the additional information needed to predict i_{n+1}, if $i_o, \ldots, i_n$ are known, or, in other words, the information loss as time passes. More precisely, the Kolmogorov entropy is defined as

$$K = \lim_{\tau\to 0} \lim_{\varepsilon\to 0} \lim_{n\to\infty} \frac{1}{n\tau} \sum_{0}^{n-1} (K_{n+1} - K_n)$$

$$= -\lim_{\tau\to 0} \lim_{\varepsilon\to 0} \lim_{n\to\infty} \frac{1}{n\tau} \sum_{i_o,\ldots,i_n} P_{i_o\ldots i_n} \log P_{i_o\ldots i_n} . \quad (1.26)$$

The limit $n\to\infty$ is first taken and later to make K independent of the chosen partition one lets $\epsilon\to 0$. Finally, the limit $\tau\to 0$ is taken only if the system is continuous. It turns out that if the motion is regular $K=0$, if it is random $K=\infty$, and if it is chaotic $0<K<\infty$.

In the case of one-dimensional systems, K coincides with the Liapunov exponent, if this is constant. For d-dimensional systems, the information loss is due only to the positive exponents. If the system is ergodic, it turns out that (Pesin, 1977 [23])

$$K = \int d^d x \, \rho(x) \sum_i \lambda_i^+ , \quad (1.27)$$

where ρ is the invariant density and λ_i^+ are the positive exponents. If it is not ergodic, but has an ergodic attractor, this formula gives the entropy of this attractor.

Time of reliable prediction.

It is easy to understand that if $K>0$, the information stored in the initial conditions will be lost after a certain finite time interval which depends on the initial precision and goes to zero if K increases. Consequently,the prediction of the state of the system will be impossible after this time. Let us consider, as an example, a discrete system in the unit interval [0,1], the initial condition x_0 being known with a precision Δx. After a time t_m, this uncertainty grows to $D=\Delta x \exp \lambda t_m$. If $D \gg 1$, the prediction has lost all its value, since nothing is known of x_m. This happens after t_m, such that

$$D = \Delta x \exp \lambda t \simeq 1 \quad \Rightarrow \quad t \simeq (1/\lambda) \log (1/\Delta x).$$

For later times only statistical predictions can be made. In the case of systems in d dimensions, the time t_m is (Farmer, 1982 [24])

$$t \simeq (1/K) \log (L/l), \tag{1.27}$$

where K is the Kolmogorov entropy, L is the characteristic size of the accessible region of phase space, and l is the initial precision. Let us stress that t_m depends linearly on K, but only logarithmically on the initial precision. If K increases by a factor z, in order to achieve the same prediction time, the initial precision must improve by a factor exp z. And to increase the time by a factor Q, the precision l has to improve by a factor exp Q.

CHAPTER 2. STRANGE ATTRACTORS

2.1 FRACTALS

The algebraic or topological dimensions of a set are always integer numbers. The definition of the former is based on the existence of basis of n vectors, such that all the elements of the set can be expressed as linear combinations of them. On the other hand, a subset has topological dimension 1 if it separates in two disconnected parts upon the elimination of one or several elements; it has topological dimension 2 if this can be done by removing one or several sets of dimension 1, etc.

There are, however, some sets which cannot be characterized by any integer dimension. This is clearly the case of the Cantor set, represented in Figure 2.1. It is certainly not a line of dimension 1, but neither it is a discrete set of points with zero dimension: it looks like something intermediate. This suggests that it is convenient to generalize the idea of dimension, so that noninteger values are admissible. In fact, this was already done by Hausdorff and further developed by Besicovitch around 1920.

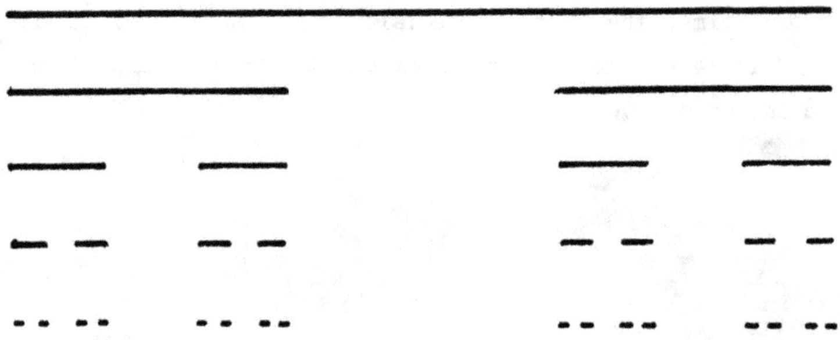

Figure 2.1. The classic Cantor set.

The general idea of dimension is not simple and can be formalized with different definitions. Perhaps the most intuitive one is what is called capacity dimension d, which is based on the following property. If we have a straight segment of length L, in order to cover it by cubes of side ϵ, we need a number of them equal to $N(\epsilon)\sim L\epsilon^{-1}$ if $\epsilon\to 0$. For a surface of area S we need $N(\epsilon)\sim S\epsilon^{-2}$. In the general case, the number $N(\epsilon)$ of n dimensional hypercubes of side ϵ, needed to cover a manifold of dimension d (n>d), is proportional to ϵ^{-d} for $\epsilon\to 0$. To be more precise, one has

$$d = -\lim_{\epsilon\to 0}[\log N(\epsilon)/\log \epsilon]. \tag{2.1}$$

If we apply this definition to general sets, it happens that d is not integer in many cases. Sets with noninteger dimension are called "fractals" [1,2]. Let us examine the Cantor set of Figure 2.1. If we take $\epsilon=3^k$, $k\to\infty$, it is clear that $N(\epsilon)=2^k$ (if k=1, we need 2 intervals; if k=2, 4 of them are required, etc.). This implies

$$d = \lim_{k\to\infty}[\log 2^k/\log 3^k] = \log 2/\log 3 \simeq 0.6309.$$

As we see, from this point of view this Cantor set is intermediate between a discrete set of points (zero dimension) and a segment (dimension 1), which is certainly interesting. Sets of this type are called dusts by Mandelbrot.

The original concept of dimension proposed by Hausdorff in 1919 is somewhat more complicated. Let a set be in an n-dimensional Euclidean space and consider its covering with n-dimensional hypercubes of variable side ϵ_k. Define the quantity $l_d(\epsilon)$ as

$$l_d(\epsilon) = \inf(\epsilon_k{}^d),$$

where $\epsilon_k \leqslant \epsilon$. Now let l_d be

$$l_d = \lim_{\epsilon\to 0} l_d(\epsilon).$$

Hausdorff showed that there exists a critical value of d above which $l_d=0$ and below which $l_d=\infty$ and took this value $d=d_H$ as the dimension of the set. It is easy to show that $d \geqslant d_H$. But while there are sets for which d and d_H are different (e. g., the harmonic sequence 1/n), it is believed that the fractals which arise in dynamical systems have $d=d_H$. For this reason it is not uncommon to refer to (2.1) as the Hausdorff dimension. Sometimes it is also called the box-counting dimension, since its calculation proceeds through the partition of the phase space into boxes, counting how many of them are occupied by points of the set. Unfortunately, when d>2 the method is very difficult and time consuming.

Another useful definition, and of easier application, is that of pointwise dimension. Let a ball of radius r, centered at a point x of the set, contain a number N(r) of points of the set that is proportional to r when $r\to 0$, or in other words let

$$\nu(x) = \lim_{r\to 0} [\log N(r)/\log r].$$

If $\nu(x)=\nu$ is independent of x, it is called the pointwise dimension. Typically, although not always, $\nu < d$.

2.2 STRANGE ATTRACTORS: THE HENON MAP AND THE LORENZ MODEL

Many dissipative dynamical systems have the property that the trajectories tend asymptotically in time to a bounded region of phase space, called an attractor, if they begin at t=0 in the so-called basin of attraction. If this is the case, the attractor is the relevant part of phase space, since the motion always ends up in it, after a sufficient time interval. It is then said that the system forgets the initial conditions as, independently of their values, provided only that they are in the basin of attraction, it always ends in the attractor.

Several remarks must be made. First, a system may have more than one attractor. Second, as all the points in the attraction basin go to the attractor, the evolution contracts volumes in phase space. For this reason conservative systems cannot have attractors. In the dissipative case, on the other hand, the volume is contracted at each step or at any time and they may have attractors which necessarily have zero volume, since the basin is only contracted to them in the limit $t\to\infty$. Let us consider some examples of attractors(see Figure 2.2):

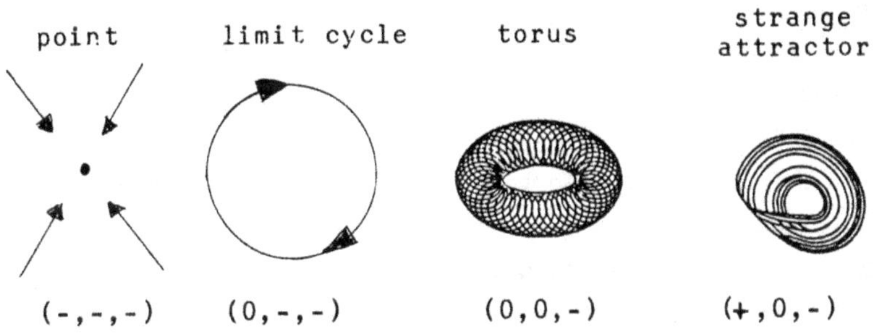

Figure 2.2. Some attractors.

(i) The damped harmonic oscillator

The equation of motion is

$$\ddot{x} + 2\gamma\dot{x} + \omega^2 x = 0 \,,$$
$$\gamma > 0 \,.$$

As is well known, all the trajectories tend to the solution $x=0, \dot{x}=0$. Therefore, the attractor is a point in phase space.

(ii) The van der Pol oscillator

This nonlinear oscillator is described by the equation

$$\ddot{x} - (\epsilon - x^2)\dot{x} + x = 0,$$
$$\epsilon > 0.$$

This is analogous to the previous case, but the effective damping coefficient $-(\epsilon - x^2)/2$ is negative for small x and positive for large x. Consequently, the equilibrium point $x=\dot{x}=0$ is an unstable repellor, since small displacements tend to increase, while for large values of x the motion is damped. As a compromise between these two tendencies, there is a periodic solution, represented by a closed curve in phase space, which attracts all the motions. If $\epsilon \ll 1$, this attractor is a circle of radius $2\sqrt{\epsilon}$. Closed curves of this type which attract all trajectories in a basin are called limit cycles (or limit cycles of Poincaré) and represent stable oscillations which build up and sustain themselves, after any perturbation, in the absence of external forces. For this reason, they are called self-excited or self-sustained oscillations.

Van der Pol proposed this equation to account for the behaviour of a triode.

(iii) Torus T^2.

The attractor can also be a torus, if the phase space has more than two dimensions.

If we consider attractors of the previous types in a phase space of three dimensions, the Liapunov exponents would be: (i) for a point (-,-,-); (ii) for a limit cycle (0,-,-); (iii) for a torus (0,0,-), where - and 0 mean negative and null value, respectively.

To end this section it is convenient to mention an important restriction on the form of an attractor in two dimensions. The well known Poincaré-Bendixson theorem states that, under very general conditions on the equations of the system, attractors in two-dimensional systems can be only of two kinds: points and limit cycles. This implies that there can be no chaotic flow in a bounded region of two-dimensional space.

In the preceding examples, the motion is regular on the attractor, as in the case of flows in 0, 1, or 2 dimensions it can not be otherwise. But if the dimension of the attractor is higher than 2 in the continuous case(or always in the discrete one), it may happen that it attracts all the trajectories in a basin and is therefore a stable set, but that the motion is unstable in it. In these cases, one speaks of strange attractors (SA), which are objects of great complexity that combine stability (since all close motions tend to them) with instability (since the motion is chaotic on them) and with the butterfly effect (strong sensitivity to initial conditions).

Definition of Strange Attractor

The idea of strange attractor is difficult to understand in an intuitive way, this explaining why it was not discovered until 1971, when it was proposed by Ruelle and Takens.[3,4] Also for this reason, there is not a universally accepted formal definition. The main characteristics of a SA are:

(i) It is an attractor, that is a bounded region of phase space towards which all the trajectories, which at t=0 are in the so called basin of attraction, tend for long enough times. It must also be indecomposable, which means that the motion must be ergodic in it.

(ii) In the attractor there is sensitivity to initial conditions (butterfly effect). This means that close trajectories separate exponentially and that one of the Liapunov exponents and the Kolmogorov entropy are positive.

(iii) It is structurally stable and generic. This means that, if the second members of the equation of the system are submitted to small changes, it does not disappear and its characteristics, for instance its Hausdorff dimension, change in a continuous way.

(iv) Frequently, SA are defined by their most amazing property, which is their fractal Hausdorff dimension. In fact, all the SA found so far have noninteger dimension. But, it must be said that it is not clear whether there are SAs with integer dimension or whether all of them are fractals and this property can be deduced from (i),(ii) and (iii).

Properties (i) and (ii) imply that SA combine transversal contraction with longitudinal dilatation. Consequently, if the flow is three-dimensional, the Liapunov exponents are (-,0,+), as it is indicated in Figure 2.2. To have a bounded structure with these properties it is necessary that it folds onto itself an infinite number of times;the first stages of this process are shown in Figure 2.3 {Bergé et al., 1984}. This gives another example of the stretching and folding mechanism which we found in Chapter 1 in the case of the Bernouilli shift and which seems to be an essential ingredient of chaotic behaviour.

Strange attractors, as any other attractors, only appear in dissipative systems, in which there is contraction of the phase space,

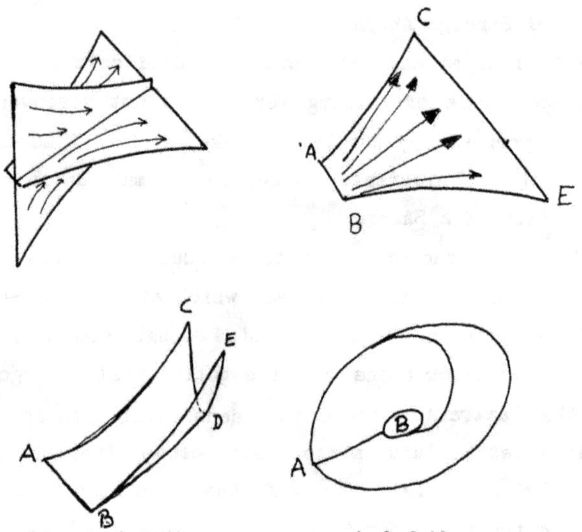

Figure 2.3. Stretching and folding in the construction of a SA.

so that their volume is zero. In the continuous case, one and two-dimensional attractors must be nonstrange according to the above mentioned Poincaré-Bendixson theorem, so that SAs must have more than two dimensions. In the discrete case there is no such restriction and they may have any dimension.

Let us now consider two classic examples of strange attractors which appear in the Hénon map, a discrete system in two dimensions, and in the Lorenz model, a continuous three-dimensional flow.

The Hénon Map

Hénon (1976)[5] proposed the following dynamical system in the plane:

$$x_{n+1} = 1 - ax_n^2 + y_n,$$
$$y_{n+1} = bx_n. \quad (2.2)$$

It is a dissipative system which contracts the area if $|b| < 1$, since the change of the area in each iteration is given by the Jacobian

$$J = \frac{\partial(x_{n+1}, y_{n+1})}{\partial(x_n, y_n)} = \begin{vmatrix} -2ax_n & 1 \\ b & 0 \end{vmatrix} = -b. \qquad (2.3)$$

It is convenient to take for b a value smaller enough than unity in order that the contraction is appreciable, but not so small that it is too large. The values a =1.4; b =0.3 have been carefully studied. It then happens that the motion either goes to infinity or to a strange attractor which can be seen in Figure 2.4, as obtained by iterating an initial condition 10000 times. Figures 2.4 b, c, and d are enlargements of the small squares in the preceding panels. It must be said that, in general, the fourth iterate cannot

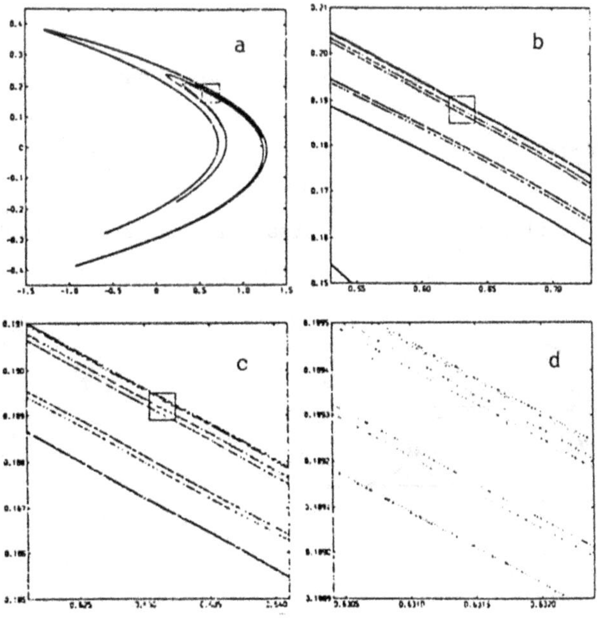

Figure 2.4. The Hénon attractor for a=1.4, b=0.3 (Hénon, 1976 [5]).

be distinguished from the attractor at the scale of Figure 2.4a. As we can see, the Hénon attractor is inhomogeneous and can be considered as the product of a curve and a set of Cantor type. Moreover, it shows scale invariance. Its capacity dimension is 1.26, as has been calculated by covering it with squares of side ϵ and using (2.1).

If we draw a circle of radius r, centered at one point of the attractor, we can see that it is transformed into an ellipse of semi-major axis $\lambda_1 r$, directed along the attractor, and semi-minor axis $\lambda_2 r$, with $\lambda_1 > 1$, $\lambda_2 < 1$ and $\lambda_1 . \lambda_2 = 0.3$.

The Lorenz Model

We have already seen how the Lorenz model (1963)[6)] played a very important role in the study of chaos. It is given by the equations

$$\begin{aligned} \dot{X} &= -\sigma X + \sigma Y, \\ \dot{Y} &= -XZ + rX - Y, \\ \dot{Z} &= XY - bZ. \end{aligned} \tag{2.4}$$

It is clear that the volume in phase space changes according to

$$dV/dt = -(\sigma+1+b)V, \qquad V(t) = V(0) \exp[-(\sigma+1+b)t]. \tag{2.5}$$

As σ and b are always positive, phase space volumes shrink and the system is dissipative.

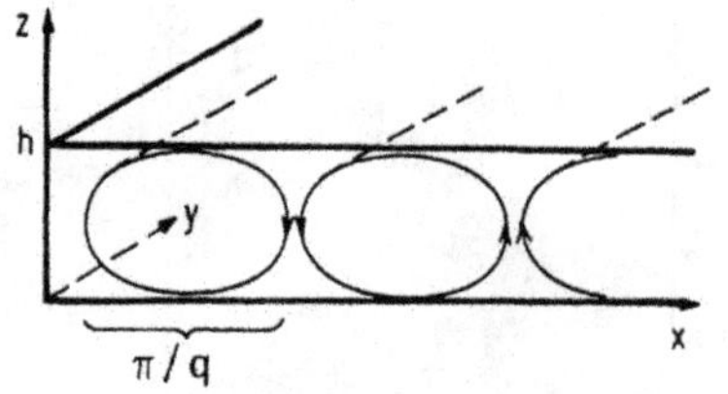

Figure 2.5. The Rayleigh-Bénard convection.

Let us see how equations (2.4) arise. First of all, let us consider a Rayleigh-Bénard convection experiment, as it is shown in Figure 2.5. After some approximations, the equations which govern the flow take the form ({Bergé et al., 1984}, {Schuster, 1984}).

- *Navier-Stokes equation*

$$(\partial\vec{v}/\partial t + \vec{v}.\vec{\nabla}\vec{v})/Pr = -\vec{\nabla}p + \theta\vec{e} + \nabla^2\vec{v}, \tag{2.6}$$

- *Continuity equation for an incompressible fluid*

$$\vec{\nabla}.\vec{v} = 0, \tag{2.7}$$

- *Equation of heat propagation*

$$\partial\theta/\partial t + \vec{v}.\vec{\nabla}\theta = Ra\,\vec{e}\vec{v} + \nabla^2\theta, \tag{2.8}$$

where v and p are the velocity and the pressure of the fuid, e is a unit vertical vector which indicates the direction of the gravity, Pr is the Prandtl number, equal to the kinematic viscosity divided by the thermal diffusivity, Ra is the Rayleigh number (1.5), and θ is the deviation of the temperature from the linear profile

$$\theta = T - T_o - \Delta T + \Delta T.z/h. \tag{2.9}$$

(2.6), (2.7), and (2.8) are a set of highly nonlinear partial differential equations the solution of which is very difficult. Lorenz obtained (2.4) as an approximation, valid in the first phase of the convection. In so doing, he proceeded in the following way. He looked for solutions independent of y, which certainly exist, as the system is invariant under the translation $y \to y+\text{constant}$. He could therefore assume that $\vec{v}=(u(x,z,t),0,w(x,z,t))$, $\theta=\theta(x,z,t)$. The continuity equation implies, then, that there is a function $\phi(x,z,t)$, such that

$$u=-\partial\phi/\partial z \quad , \quad w=\partial\phi/\partial x, \tag{2.10}$$

Lorenz expanded θ and ϕ in Fourier series and kept only the first few terms. More precisely, he took

$$\begin{aligned} \phi(x,z,t) &= \phi_1(t)\sin(\pi z/h)\sin(qx), \\ \theta(x,z,t) &= \theta_1(t)\sin(\pi z/h)\cos(qx) + \theta_2(t)\sin(2\pi z/h), \end{aligned} \tag{2.11}$$

and substituted this in (2.6)-(2.8), thus obtaining three ordinary differential equations for ϕ_1, θ_1 and θ_2, which after writing

$$\phi_1 = \frac{\pi^2+q^2}{\pi q}\sqrt{2}\,X(t) \;,\; \theta_1 = \frac{(\pi^2+q^2)^3}{\pi q^2}\sqrt{2}\,Y(t),\; \theta_2 = -\frac{(\pi^2+q^2)^3}{\pi q^2}Z(t),$$

$$r = \frac{Ra\, q^2}{(\pi^2+q^2)^3},\quad b = \frac{4\pi^2}{(\pi^2+q^2)} \quad , \quad \sigma = Pr \;, \tag{2.12}$$

take the form (2.4).

Two remarks are in order: (i) the quantities X,Y,Z are not spatial coordinates, but represent the velocity and the temperature of the fluid through (2.10), (2.11), and (2.12); (ii) The Lorenz model describes Rayleigh-Bénard convection only in the proximity of the transition from thermal conduction to convection, when there are only simple rolls. For this reason, the chaotic behaviour which is observed in the experiments for high values of Ra has nothing to do with (2.4). For more details, see {Bergé et al., 1984} or {Schuster, 1984}.

The Lorenz model has been investigated in great detail by numerical means since there are no analytic solutions. Frequently, the values b=8/3 and σ=10 are taken. In that case one has V(t)=V(0)exp(-41t/3). The third parameter, r, is the control parameter and is proportional to the difference of temperature ΔT between the two surfaces of the fluid. The main properties of (3.4) turn out to be the following:

(i) If $0 < r \leq 1$, there is only one stationary solution X=Y=Z=0, which happens to be stable. It represents a state in which the fluid is at rest and the heat transmission proceeds by conduction.

(ii) If r passes through the value 1, the preceding solution becomes unstable and two new stable sattionary states C, C' are created at

$$X=Y= \pm \sqrt{[b(r-1)]}, \quad Z= r-1 \tag{2.13}$$

by a pitchfork bifurcation in which a stable point is transformed into an unstable plus two stable points. All the initial conditions tend asymptotically to C or C', which represent roll convection states, each one in with a sense of rotation.

(iii) At r=24.06 a strange attractor is generated which coexists with C and C' up to r=24.74, when C and C' lose their stability. For higher values of r, up to r=30.1, all the motions go to the strange attractor.

(iv) After r=30.1, the r dependence is very complex with alternation of strange and periodic attractors.

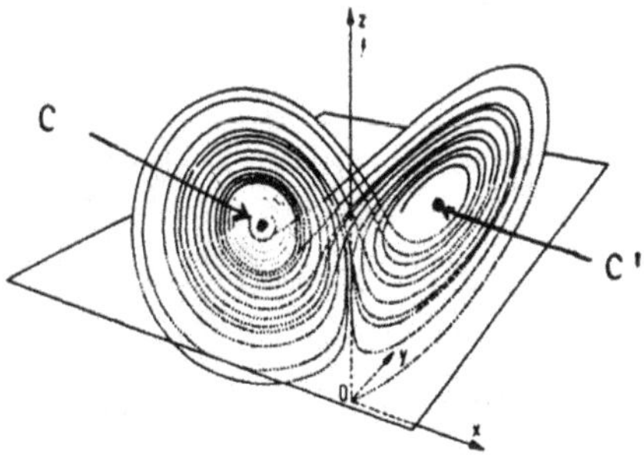

Figure 2.6. The Lorenz attractor. (after O.E.Lanford, 1977 [7]).

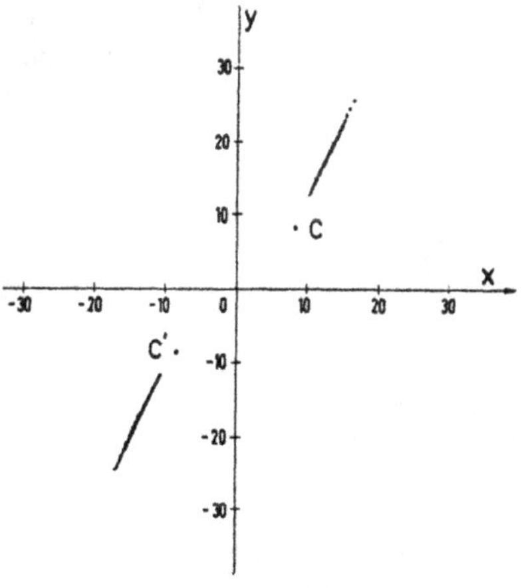

Figure 2.7. Poincaré map in the Lorenz model (after (Bergé et al., 1984)).

The most characteristic structure of the Lorenz model is, of course, the strange attractor; it appears in Figure 2.6, where the two points of unstable equilibrium are clearly seen (for r=28), as well as how the motion in the attactor consists of oscillations around them. To understand better its structure, it is helpful to examine a Poincaré section, for which the plane Z=r-1 is specially adequate, as it contains the points C and C'. As we can see in Figure 2.7, this section seems to be formed by two segments which correspond to rotations around C and C'. This could lead to the erroneous conclusion that the SA is a two-dimensional surface, but it turns out that this is impossible. Because if we label a solution by the sequence $(n_1, n'_1, n_2, n'_2, \ldots, n_k, n_k, \ldots)$, which means that it turns n_1 times around C, n_1 around C', ..., n_k around C, etc., it happens that this sequence is chaotic, as it does not follow any recognizable pattern (Figure 2.8).

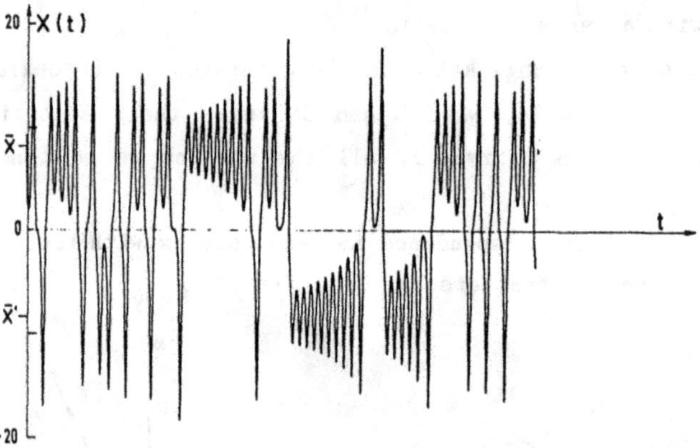

Figure 2.8. Coordinate X(t) in the Lorenz model (after {Bergé et al, 1984}).

As all the turns around C (or C') are given in the same sense, it is impossible that the attractor is a surface, since the trajectories would cut each other. If a finer calculation is made, it can be seen that it is formed by a large number (in fact it is infinite) of very tight sheets, something similar to the product of a surface and a Cantor set. Its dimension can be calculated by making a partition of the phase space into cubic cells of side ϵ and counting how many are occupied by its points , making use of (2.1). The result, for r=28, is 2.06. This corresponds to more than a surface, but not to much more, this being due to the strong contraction of volumes, previously noted. The fact that the dimension of the attractor is close to that of a surface makes possible to study its properties in an approximate way by means of a first return map. This was already done by Lorenz himself when he studied the succesive maxima of Z, each representing

Z_{k+1} as a function of the preceding one, Z_k. As the maxima of Z satisfy XY-bZ=0, this diagram coincides with the Poincaré map on the surface defined by this equation. In this way, the three-dimensional flow is reduced to a one-dimensional aplication of type $x \to f(x)$. As we have already remarked, the instabilities in this applications correspond to values of the derivative $f'(x)$ with modulus greater than one. And, as we can see in Figure 2.9, this is always the case in the range $25 < r < 30$. Correspondingly, trajectories separate exponentially in time, as is shown in Figure 2.10, where the saturation at long times is due to the fact that this separation becomes of the same order as the attractor size. By measuring the slope of the curve, it is easy to estimate the greatest Liapunov exponent to be approximately equal to 0.9, clearly greater than 0.

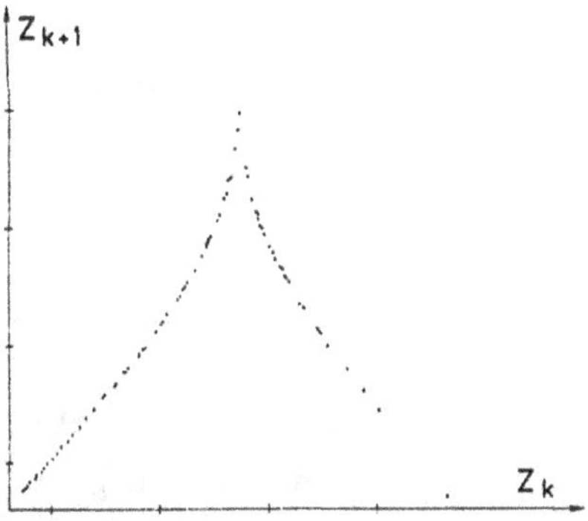

Figure 2.9. Return map in the Lorenz model (after (Bergé et al., 1984)).

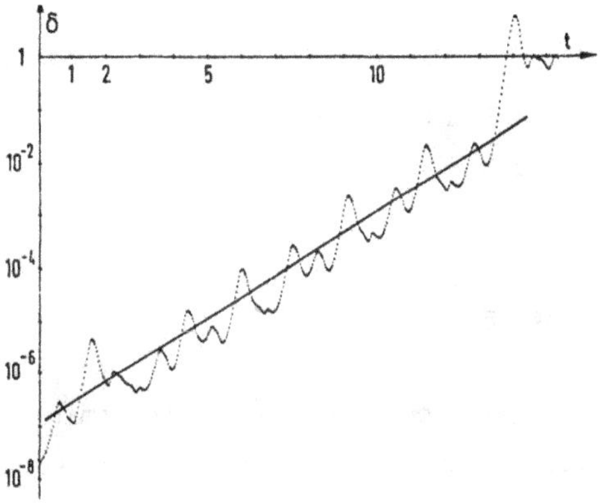

Figure 2.10. Separation of two close trajectories in the Lorenz model (after (Bergé et al., 1984)).

More on the Dimension of a Strange Attractor

We have given two different definitions of fractal dimension, the capacity and the pointwise dimension. In order to systematize this concept, Grassberger, Hentschel, and Procaccia[8,10] defined an infinite number of dimensions D_0, D_1, D_2, ..., etc., as follows. Take a sequence of points on the attractor $x_k = x(k\tau)$, k=0,1, ... ,M and divide the phase space into cubic cells of side ϵ. Let $p_i = \lim (M_i/M)$, $M \to \infty$, M_i being the number of points in cell i, and let $N(\epsilon)$ the number of cells which are occupied by some point. Then, we define the dimension D_k as

$$D_k = \lim_{\epsilon \to 0} \frac{1}{k-1} \frac{\log(\Sigma p_l^k)}{\log \epsilon}, \quad k=0,1,2,\ldots \quad . \tag{2.15}$$

Let us consider the lowest values of k. First, for k=0 we have (taking the limit $k \to 0$)

$$D_0 = -\lim_{\epsilon \to 0} [\log N(\epsilon)/\log \epsilon], \tag{2.16}$$

which is the same as the capacity dimension.

If $k \to 1$,

$$D_1 = -\lim_{\epsilon \to 0} [S(\epsilon)/\log \epsilon], \tag{2.17}$$

where

$$S(\epsilon) = -\Sigma\, p_k \log p_k \;. \tag{2.18}$$

As we have seen, $S(\epsilon)$ is the information gain if the cell in which the system is at time t is determined, so that D_1 expresses how this gain increases if $\epsilon \to 0$, and is accordingly called information dimension. If the attractor is homogeneous, $p_l = 1/N(\epsilon)$, from which $S(\epsilon) = \log N(\epsilon)$ and $D_0 = D_1$. The difference between D_0 and D_1 gives thus an indication of its inhomogeneity. Moreover, it is easy to show that $S(\epsilon) \leqslant \log N(\epsilon)$, that is, $D_0 \leqslant D_1$. In fact, it can be proved that $D_{k'} \leqslant D_k$ if $k' > k$.

If k=2, we obtain a dimension of great physical significance. It turns out that D_0 and D_1 are difficult to compute, specially if they are big. Fortunately enough, it is far simpler to determine D_2:

$$D_2 = \lim_{\epsilon \to 0} [\log(\Sigma\, p_k^2)/\log \epsilon], \tag{2.19}$$

since it can be deduced from the so-called correlation integral

$$C(\epsilon) = \lim_{N\to\infty} (1/N^2) \sum_{i,j} \theta(\epsilon - |\vec{x}_i - \vec{x}_j|), \qquad (2.20)$$

where θ is the step function, because $\Sigma\ p_k^2$ is the probability that two points of the attractor are in the same cell and this, in its turn, is approximately the same as the probability that two points are separated by a distance smaller than ϵ, a quantity precisely given by the correlation integral (2.20). As is seen, D_2 coincides with the pointwise dimension; it is frequently called correlation dimension also. From the numerical point of view, it is far simpler to calculate $C(\epsilon)$ from (2.20), rather than D_o, from (2.16). One has then,

$$\log C(\epsilon) = D_2 \log \epsilon, \qquad (2.21)$$

so that D_2 is the slope in the diagram log $C(\epsilon)$ vs log ϵ (see Figure 2.11). It must be stressed that D_2 gives a lower bound of D_o. For instance, in the Hénon attractor the values of these dimensions are D_o =1.26 and D_2=1.21 (for a=1.4, b=0.3).

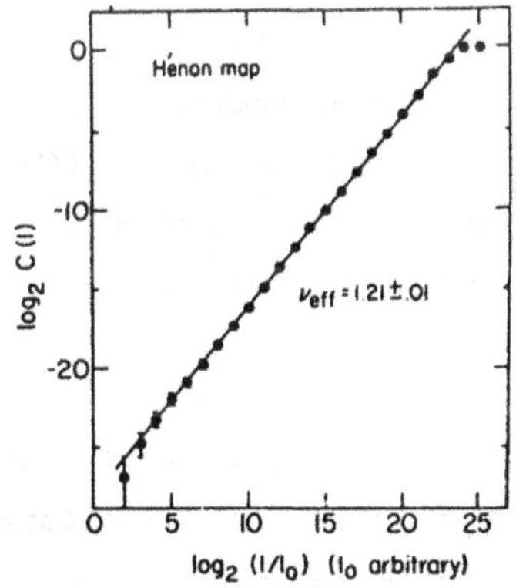

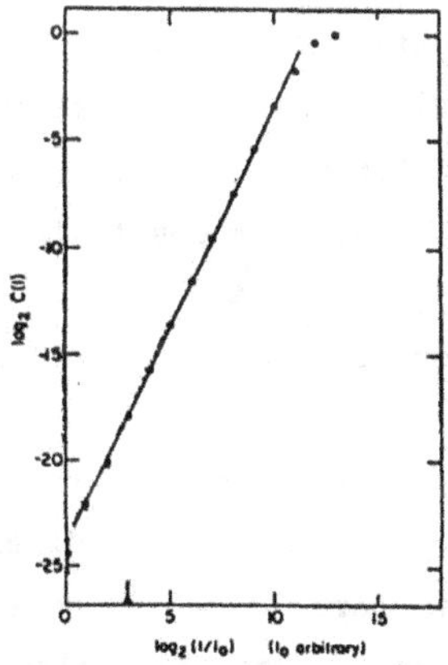

Figure 2.11. LogC(ϵ) versus logϵ in the Hénon map(left) and in the Lorenz model (right)(after Grassberger and Procaccia, 1983 [9]).

Takens' Theorem

Frequently and for practical reasons, only one of the coordinates is measured. It might, thus, seem to be impossible to investigate many of the properties of the flow in phase space and, in particular, the structure of strange attractors. It is fortunate, however, that certain aspects of the flow can be reconstructed from the sequence of the values of a single variable. Let us take, as an example, the Lorenz model. If we only know the time series of the velocity, which is simpler to measure than the temperature, we can try to reproduce the three-dimensional time evolution of $X(t)\equiv(X(t),Y(t),Z(t))$ by studying instead the vector $\chi(t)\equiv(X(t),X(t+\tau),X(t+2\tau))$, where τ is a certain delay, the time interval between successive measurements, for instance. It can be expected that there is some relation between both time evolutions, which would make easier the study of the model. It turns out that this is the case, as is shown in a theorem due to Takens (1981) [11], which states:

If $\dot{x}=F(x)$ is a flow in d dimensions, the transformation

$X(t) \rightarrow \chi(t)\equiv(x_j(t),\ x_j(t+\tau),\dots,\ x_j(t+(2d+1)\tau))$

where x_j is an arbitrary component of $X(t)$, provides a smooth embedding for this flow, in the sense that distances in $X(t)$ and $\chi(t)$ have a ratio which is uniformly bounded and bounded away from zero. Consequently, the metric properties in both spaces (the d-dimensional $X(t)$ and the (2d+1)-dimensional $\chi(t)$) are the same.

It turns out that, in many cases it is not necessary to take as many as (2d+1) components for the vector χ. If we take $\chi\equiv(X(t),X(t+\tau),\dots,X(t+(p-1)\tau))$, the minimum dimension p above which D_2 of a strange attractor no longer changes is called its embedding dimension.

The Conjecture of Kaplan and Yorke

As we have seen, the dimension of the attractor and the Liapunov exponents are closely related (see figure 2.2). If in three-dimensional space all the Liapunov exponents are negative, the dimension is 0, while, if one is zero and the other two are negative, it is 1. Kaplan and Yorke (1979) [12] conjectured that this is a particular case of a general relation which can be written as

$$D = j + \sum_{i=1}^{j} \lambda_i/|\lambda_{j+1}| \qquad (2.27)$$

where the exponents are labelled as $\lambda_1 \geq \lambda_2 \geq .. \geq \lambda_d$ and j is the largest integer for which the sum of the first j exponents is positive. This conjecture has been shown to be valid in the cases to which it has been applied, at least to a very good approximation, but a proof (or a disproof) of its validity is still lacking.

2.3 HOPF BIFURCATIONS

In some occasions, when a parameter passes through a critical value, the topological properties of the solutions in the phase space suffer a change. It is then said that a bifurcation has been produced. We will now consider the very important case of the so-called Hopf bifurcation, in which a limit cycle is generated around an equilibrium point, the stability properties of which change. As an example, let us consider the van der Pol oscillator,

$$\ddot{x} - (\epsilon - x^2)\dot{x} + x = 0. \tag{2.28}$$

If we define the energy as $E(x,x)=(\dot{x}^2+x^2)/2$, it is easy to show that its derivative with respect to time is

$$\dot{E} = (\epsilon - x^2)\dot{x}^2. \tag{2.29}$$

It is clear, then, that if $\epsilon<0$, the energy decreases along any trajectory for almost any time, from which it follows that all motions end in the stable equilibrium point (0,0). On the other hand, if $\epsilon>0$, the energy increases for small x and a limit cycle appears as a compromise between this effect and the dissipation at large values of x. To understand this phenomenon, let us consider the time average of the derivative of energy,

$$\bar{\dot{E}} = \lim_{T\to\infty} \frac{1}{T}\int_0^T \frac{dE}{dt}\, dt = \overline{\epsilon\dot{x}^2} - \overline{x^2\dot{x}^2} \tag{2.30}$$

It can be shown that, to first order in ϵ (provided that it is positive), there is a solution of the form $x=\rho \sin t$, with $\rho=2\sqrt{\epsilon}$, which, according to (2.30), has a constant average energy, because the two terms cancel each other. It is certainly a limit cycle. To summarize: If $\epsilon<0$, all trajectories are attracted to the stable equilibrium point $x=0$ ((0,0) in phase space) and there are no periodic solutions. If $\epsilon>0$, on the other hand, the equilibrium point becomes an unstable repellor and a stable limit cycle appears which attracts all the trajectories. Moreover, and this is very important, *a periodic solution is generated at the bifurcation.*

It is to be noted that a pair of conjugate eigenvalues of the matrix of the linearized differential equations of the system cross the imaginary axis as their real part change sign. In the case of the van der Pol oscillator, this matrix is

$$A = \begin{pmatrix} 0 & 1 \\ -1 & \epsilon \end{pmatrix},$$

which has the eigenvalues $\lambda=[\epsilon\pm i\sqrt{4-\epsilon^2}]/2$.

2.4 THE LANDAU ROUTE TO TURBULENCE AND THE RUELLE-TAKENS ROUTE TO CHAOS

Landau originated the idea that generically systems are regular and that turbulence arises from the complexity due to the existence of a great number of variables. Consequently, he proposed in 1944[13,14] a theory in which the transition to the turbulence proceeds through an infinite number of Hopf bifurcations, as is schematically shown in Figure 2.12a. When the control parameter surpasses the critical value R_n, a new frequency is generated as the motion changes from (n-1)-periodic to n-periodic (or, equivalently, as the trajectories which were inscribed in a (n-1)-dimensional torus become contained in a torus of n dimensions). As more and more bifurcations take place and more and more frequencies appear, the motion becomes more complex. But although the spectrum has many lines, it is discrete and can be considered as continuous only at the end of the infinite cascade.

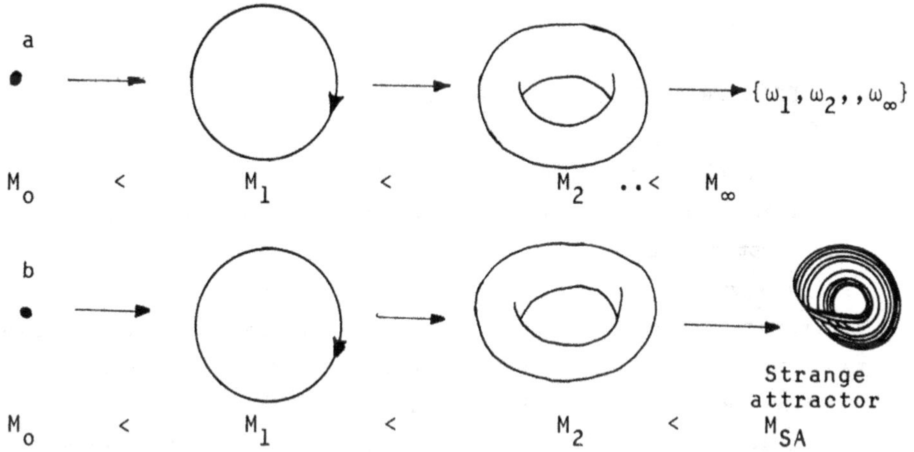

Figure 2.12. The Landau (a) and the Ruelle-Takens (b) routes.

In recent years, some systems have been found in which chaos arises directly from biperiodic motion, a transition not considered in Landau's theory. Curiously enough, this was not known in 1971 when Ruelle and Takens had a revolutionary idea, suggesting a route to chaos which is much shorter than that of Landau. They showed that after just two Hopf bifurcations, multiperiodic motion is very unstable and decays to chaotic behaviour at a strange attractor(see Figure 2.12b). It is easy to understand that chaos cannot appear until after two Hopf bifurcations, since biperiodic trajectories are situated in a bidimensional torus, in which there cannot be chaotic motion according to the Poincaré-Bendixson theorem. What is surprising is that this is enough, in the sense that, after the next bifurcation, a strange attractor has to be expected as something quite natural. To be specific, Newhouse, Ruelle, and Takens[15)] completed the first proposal by proving in 1978 that triply-periodic motion is unstable, in the sense that arbitrary small perturbations, with the only condition that they have small first and second derivatives, convert it to a chaotic flow in a strange attractor. We can thus say that strange attractors are generic and triply-periodic flows highly improbable.[16,17]

2.5 EXPERIMENTAL EVIDENCE OF THE RUELLE-TAKENS ROUTE

The RT route has been observed in many situations. As an illustration, we may mention two experiments by Swinney and Gollub (1978)[18]. The first one refers to the Rayleigh-Bénard convection (Figure 1.10). For low values of the relative Rayleigh number, the motion in the attractor is simply periodic (a). If this control parameter increases the motion is successively: biperiodic with two incommensurable frequencies (b), chaotic with some lines (c), and fully chaotic (d).

The Taylor instability occurs in the Couette flow, in which a fluid layer is situated bewteen an inner cylinder rotating with angular velocity Ω and a stationary outer one (Figure 1.10). For small Ω, the angular momentum is transported from one cylinder to the other by viscosity and the flow is very regular (a). After a critical value Ω', this regime becomes unstable and annular convection rolls appear in a clearly biperiodic motion (b). At still higher values chaos appears.

In these two examples, the transition to chaos follows clearly the Newhouse-Ruelle-Takens route.

It must be mentioned that Grebogi, Ott, and Yorke[19] have shown that triply-periodic flow can still persist if the perturbations have small derivatives of all orders, and not only first and second ones as NRT had assumed in their theorem. Their work was numerical, but the existence of triply-periodic flows has been observed in some cases. For instance in a Rayleigh-Bénard experiment with mercury by Libchaber, Fauve and Laroche (1983)[20] and in the voltage spectrum of a ferroelectric crystal of Barium Sodium Niobate (BSN) by Martin, Leber, and Martienssen (1983)[21].

CHAPTER 3 THE SUBHARMONIC CASCADE

3.1 INTRODUCTION

In the Feigenbaum route or scenario, one of the three classic roads, the transition to chaos proceeds in the following way:

A dynamical system has a stable periodic solution with period T for certain values of a parameter μ, on which it depends. If μ is varied, it turns out that

(i) When it reaches a certain value μ_1, there is a bifurcation at which the periodic orbit loses its stability and a new stable solution with period 2T is generated. This is followed by an infinite sequence of similar period doubling bifurcations $\mu=\mu_k$, $k=2,3,\ldots,\infty$, at which solutions with periods $2^{k-1}T$ loose their stability while new stable ones with periods 2^kT are created.

(ii) The sequence μ_k approaches its limit at the rate

$$\mu_K = \mu_\infty - \text{constant } \delta^{-k},$$

where δ has always the universal value 4.66920....

(iii) After μ_∞, the motion is mostly chaotic.

In this chapter, the main characteristics of this so-called subharmonic or Feigenbaum sequence will be described.

3.2 THE PERIOD DOUBLING PHENOMENON

Let $f(x,\mu)$ be a unidimensional map depending on the parameter μ. Its iteration defines a discrete dynamical system :

$$x_{n+1} = f(x_n,\mu) \ . \tag{3.1}$$

Let x^* be a fixed point depending on μ. If $x=x^*+y$, $y<<1$, the linearization around x^* gives

$$y_{n+1} = f'(x^*,\mu)\, y_n = a\, y_n ,$$

the solution of which is $y_n=y_o a^n$, x^* being stable if $|f'(x^*,\mu)|<1$. If this is the case, it may happen that, after a change in μ, the value of $|f'(x)|$ will become greater than one, with the corresponding loss of the stability of x^*. This may occur in two different ways, according to whether f' crosses the values +1 or -1. Let us examine in detail these two cases.

(i) $f'(x^*,\mu) = -1-\epsilon$,

where $\epsilon=\epsilon(\mu)$ passes from negative to positive values (Figure 3.1): The second iterate of f, $f^2(x,\mu)=f[f(x,\mu),\mu]$, defines another dynamical system,

$$x_{n+1} = f^2(x_n,\mu) ,$$

which also has x^* as a fixed point. Moreover, since

$$f^{2\prime}(x^*,\mu) = \{f'(x^*,\mu)\}^2 ,$$

its stability properties with respect to f and f^2 are the same (if $|f'|<1$, then $|f^{2\prime}|<1$ and conversely). As a clear consequence, if x^* loses its stability through -1 with respect to f, it does the same thingthrough +1 with respect to f^2. In other words, f^2 is in the following case.

(ii) $f'(x^*,\mu) = 1+\epsilon$,

$\epsilon=\epsilon(\lambda)$ increasing from negative to positive values: The case we are considering is represented in Figure 3.2, where one can see that $f'''(x^*)<0$ and that $f''(x^*)=0$ at the bifurcation, since f' has a maximum. We may write the map, up to cubic terms, as

$$y_{n+1} = f'(x^*,\mu)\, y_n + \frac{1}{6} f'''(x^*,\mu)\, y_n^3,$$

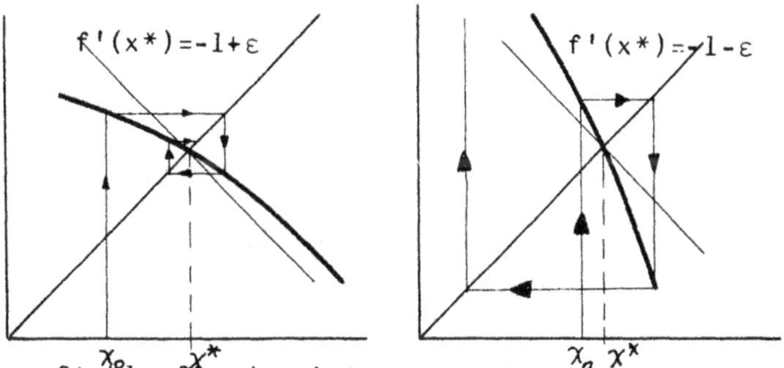

Stable fixed point   Unstable fixed point

Figure 3.1. Bifurcation through f'=-1.

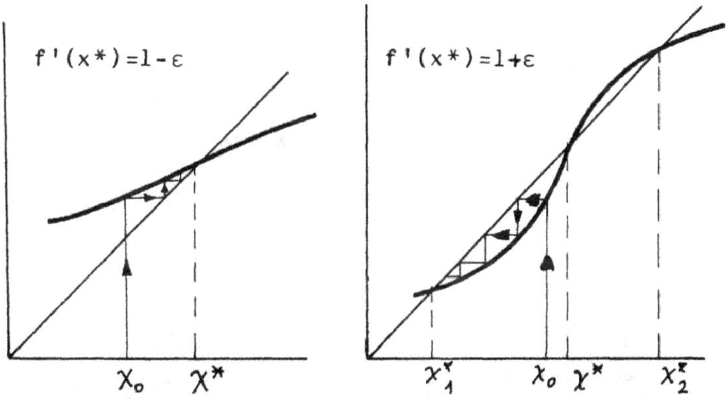

A stable fixed point   An unstable and two stable fixed points

Figure 3.2. Bifurcation through f'=+1.

the fixed points being

$$y_n = 0, \quad y_n = \pm\sqrt{-6\,\frac{f'-1}{f'''}}$$

(for μ close enough to its bifurcation value, the term in f'' can always be neglected if $f'''<-\rho<0$). The first one is x*, while the other two exist only for f'>1 (after the bifurcation) and correspond to

$$x_n = x^*_{1,2} = x^* \pm\sqrt{\frac{-6\varepsilon}{f'''}} \ .$$

Moreover

$$f'(x^*) = 1 + \varepsilon,$$

$$f'(x^*_{1,2}) = f'(x^*) + \frac{1}{2} f'''(x^*) \{\pm\sqrt{\frac{-6\varepsilon}{f'''}}\}^2 = 1 - 2\varepsilon,$$

so that after the bifurcation x^* is unstable, while x^*_1 and x^*_2 are stable.

Going back now to the case (i), in which f' crosses the value -1, we see that, if $f'''(x^*)<0$, then

(a) if x^* is a fixed point of f, it is also a fixed point of f^2 and of all the higher order iterates f^k.

(b) when x^* looses its stability with respect to f, it becomes simultaneously unstable with respect to f^2 and to all the iterates f^k.

(c) at the same time two new stable fixed points of f^2, x^*_1 and x^*_2 are generated. This pair is called a period two attractor of f, since the sequence $x^*_1, x^*_2, x^*_1, x^*_2, \ldots$, is a trajectory of f, to which all near orbits tend asymptotically. This is because, as

$$f^2(f(x^*_1)) = f(f^2(x^*_1)) = f(x^*_1)$$

$f(x^*_1)$ is a fixed point of f^2 and it can only be $x^*_2 = f(x^*_1)$ (conversely $x^*_1 = f(x^*_2)$). Moreover, the derivative

$$f^{2\,\prime}(x^*_1) = f'(x^*_1)f'(x^*_2) = f^{2\,\prime}(x^*_2) \tag{3.2}$$

has a modulus smaller than one. The bifurcation can be represented by the diagram at right and is called accordingly pitchfork bifurcation

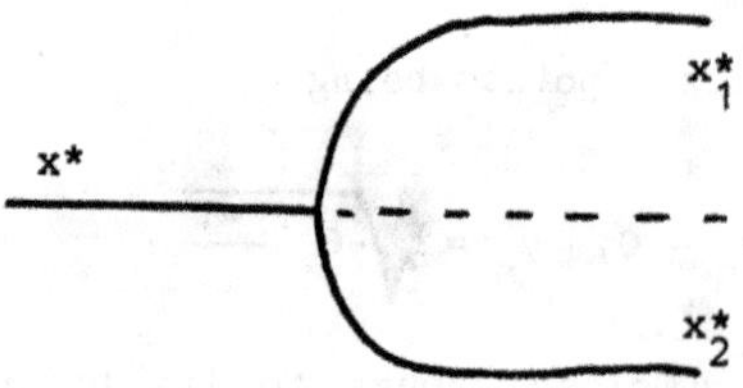

(d) if we keep varying μ, x^*_1 and x^*_2 become simultaneously unstable because of (3.2). Each one gives rise then to two stable fixed points of $f^4 = f^2(f^2)$, that is, to a period 4-attractor of f.

A more detailed analysis (Collet and Eckmann, 1980)[1] shows that if the function $f(x,\mu)$ is a map of class C^1 of the unit interval into itself, has a maximum and is monotone at both sides of it, and has a negative Schwartz derivative Sf, that is, if

$$Sf = f'''/f' - (3/2)\,(f''/f')^2 \sim \frac{d^2}{dx^2}\,(f')^{-1/2} < 0,$$

then the variation of the parameter μ produces an infinite sequence of pitchfork bifurcations which generate attractors of period 2^k. This is precisely the so called Feigenbaum cascade. Clearly, if f is a map on a general finite interval, the same conclusion applies, as can be seen by a simple change of variable.

The case of several dimensions is, in principle, more complex. However, we can reduce it to a lower dimension number and even to just one, by means of suitable Poincaré maps, for instance. We may also take the restriction of the system to an eigendirection of the evolution operator. Curiously enough, this allows one to find in many cases the same phenomenon of the infinite sequence of period doublings, chaos appearing at the end. Moreover, and this is really surprising, several quantitative properties are the same as those appearing in unidimensional maps.

3.3 THE LOGISTIC MAP

To illustrate the Feigenbaum cascade, the so-called logistic map of the unit interval into itself,

$$f_\mu(x) = \mu x(1-x)\ , \quad 0 < x < 1, \quad 0 < \mu < 4\ , \tag{3.3}$$

and the associated dynamical system

$$x_{n+1} = f_\mu(x_n) = \mu x_n(1-x_n) \tag{3.4}$$

are frequently used. We can see in Figure 3.3 the shape of $f_\mu(x)$. For convenience, the subscript μ will be omitted if there is no risk of confusion.

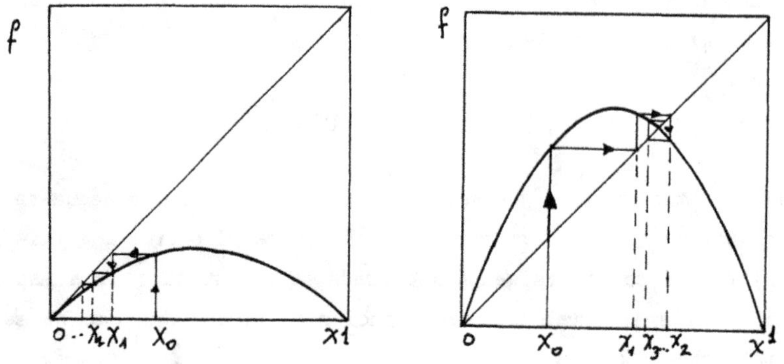

(a) (b)

Figure 3.3. The logistic map.
(a) sequence $\{x_n\}$ for $\mu<1$; (b) same as (a) for $1<\mu<3$.

It seems that this system was first studied in 1845 by P. F. Verhulst [2] in developing a mathematical model of the population growth of a biological species in an isolated area (e.g., birds in a valley, fishes in a lake). For values of μ slightly above 1, (3.4) has the constant solution $x_n=1-1/\mu$ which, as almost all the solutions tend to it, could represent the natural population of the area. For higher values of μ, the situation is more complex, most trajectories ending in periodic solutions, a phenomenon which models the succesion of years of high and low population, frequently observed in nature and familiar to hunters and fishermen.

The logistic map appears in many different contexts, as in the study of the rotator submitted to periodic impulsive forces or the case of a savings account with self-limiting interest rate. However its main interest does not lie on its usefulness as a model of some particular phenomenon, but in the clarity and expresiveness with which it exhibits one way of transition to chaos to be found in much more complex systems.

Properties of the Sequences x_n

Let us consider (3.4) in detail. If the initial value x_0 is given, the sequence $\{x_n\}$ can be obtained by a simple geometrical construction, already used in Figures 3.1 and 3.2 and explained in

Figure 3.3. It consists in the following. First draw a vertical straight line with abscissa x_0 up to the intersection with the curve $f(x)$. Then a horizontal line from this point to the intersection with the bisector of the first quadrant. The abscissa of this point is x_1. Repeating the process— vertical to $f(x)$, then horizontal to the bisector— it is easy to obtain $x_2, x_3, \ldots$. This sequence has been given different names: orbit, trajectory, itinerary, solution,... .

If $0<\mu<1$ (Figure 3.3a), the only fixed point is $x=0$. As $f'(0)=\mu<0$, it is stable. At $\mu=1$, there is a bifurcation in which $x=0$ becomes unstable and a new fixed point $x^*=1-1/\mu$ appears with a stability which depends on μ, since $f'(x^*)=2-\mu$.

If $1<\mu<3$ (Figure 3.3b), x^* is stable. But if μ crosses the value 3, another bifurcation takes place and, since $f'(x^*)$ passes through -1, x^* becomes unstable, a period-two attractor being generated by the previously explained process. In Figure 3.4, f and f^2 are represented for values of μ slightly below and above 3.

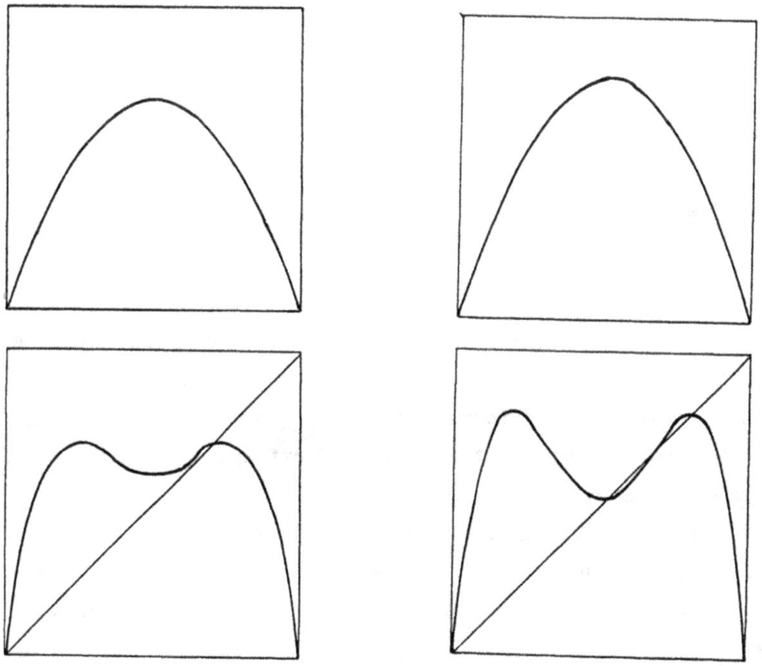

Figure 3.4. f and f^2: (a) for $\mu=2.8$ and (b) for $\mu=3.2$.

It is easy to show that f^2 has three extrema, because

$$f^{2\prime} = f'(f(x))\, f'(x)$$

is satisfied if $x = 1/2$ (as $f'(1/2)=0$) or if $f(x)=1/2$, that is if $x^*_{1,2}=f^{-1}(1/2)$. The important role of the 2-valuedness of f^{-1} appears clearly here.

If μ increases, the modulus of $f^{2\prime}(x^*_{1,2})$ reach the value -1 at $\mu=\mu_2$, with the consequent appearance of a period 4 attractor of f. For still higher values other period doublings take place, a sequence μ_k, $k=1,\ldots,\infty$ existing such that, if $\mu_{k-1}<\mu<\mu_k$, there is an attractor of period 2^{k-1} with elements x^*_i, $i=1,\ldots,2k-1$, which attracts almost all the trajectories and such that

$$f(x^*_i) = x^*_{i+1} \quad (\text{if } i = 2^{k-1}+1,\ x^*_i = x^*_1), \tag{3.5}$$

$$f^{2^{k-1}}(x^*_i) = x^*_i, \tag{3.6}$$

$$\left|\frac{d}{dx} f^{2^{k-1}}(x^*_i)\right| = \left|\prod_1^{2^{k-1}} f'(x^*_k)\right| < 1\,. \tag{3.7}$$

In Figure 3.5, the points of the attractor are represented schematically versus μ. In it, μ_k is the μ value at which a stable cycle of period 2^{k-1} bifurcates to one of period 2^k. M_k is the one for which x=1/2 belongs to the period 2^k attractor. In this case, d_k is the distance from 1/2 to the closest point in the cycle.

In spite of the simplicity of f(x), the expression of f^k becomes quickly very complex and it is necessary to resort to numerical methods. The results obtained in this way can be summarized as follows.

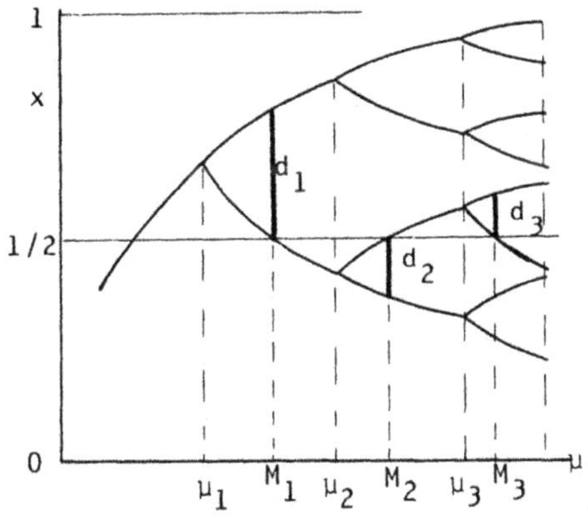

Figure 3.5. Schematic representation of the attractor versus μ in the periodic region. The distances d_n are indicated.

There are two different types of behaviour: the so-called periodic regime, corresponding to $1<\mu<\mu_\infty$, and the chaotic regime for $\mu_\infty<\mu<4$, where (see Figure 3.6)

$$\mu_\infty = \lim_{n\to\infty} \mu_\infty = 3.56994546\ldots . \qquad (3.8)$$

Periodic Regime

(a) The sequence $\{\mu_n\}$ converges to its limit as

$$\mu_n = \mu_\infty - \text{const.}\ \delta^{-n}. \qquad (3.9)$$

(b) The distances d_n satisfy

$$\frac{d_n}{d_{n+1}} = -\alpha \quad \text{for } n >> 1. \qquad (3.10)$$

(c) α and δ, the so-called Feigenbaum numbers, have the values

$$\alpha = 2.50290787\ldots,$$
$$\delta = 4.66920160\ldots, \qquad (3.11)$$

and are the same for any map with a quadratic maximum. Moreover the M_n have the same behaviour:

$$M_n = \mu_\infty - \text{const.}\ \delta^{-n} \qquad \text{for} \quad n >> 1, \qquad (3.9')$$

with the same limit for $n\to\infty$ and the same δ. One says that α and δ are universal, as they depend only of the kind of maximum of f(x). In fact, if near its maximum f(x) has the form

$$f(x) = f(x_m) + a\ (x-x_m)^z + \ldots,$$

the corresponding Feigenbaum numbers depend only on z.

Chaotic Regime

(a) For $\mu>\mu_\infty$ the behavior is chaotic for most values of μ in $[\mu_\infty, 4]$. First the iterates of $f^k(x)$ do not cover densely the whole unit interval, but fall chaotically in some subintervals which later join each other by inverse bifurcation, until at $\mu=4$ the itinerary generated by a generic x becomes dense in [0,1].

(b) There are periodic windows which are μ intervals in which the behaviour is regular, with cascades of attractors of periods p, $2p,\ldots,2^kp,\ldots(p=3,4,5,\ldots)$. The corresponding values of μ scale as (3.9) with the same δ as in the periodic region, although with a different constant and limit.

(c) There are also cascades of period triplings 3^kp and quadruplings 4^kp, and they also scale as (3.9), although with different δ. For instance, in the case of the triplings $\delta=55.247\ldots$.

(d) The order in which the periodic windows appear is also universal. In this context two windows with the same period are distinguished by the order in which the system visits the points of the attractor . The set of windows is accordingly called the universal sequence.

In the frontier between the two regimes ($\mu=\mu_\infty$) the behavior is aperiodic but not chaotic. This is because almost all the trajectories tend to an attractor in which the motion is ergodic but without sensitive dependence on initial data (the trajectories do not diverge exponentially). This attractor is a fractal set and its Hausdorff dimension is D=0.538.

In Figure 3.7 we can see the μ dependence of the attractor and the value of the Liapunov exponent λ (recall that $\lambda>0$ implies chaos).

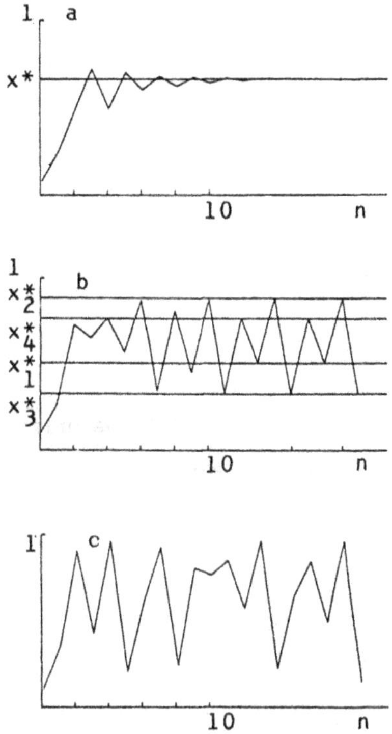

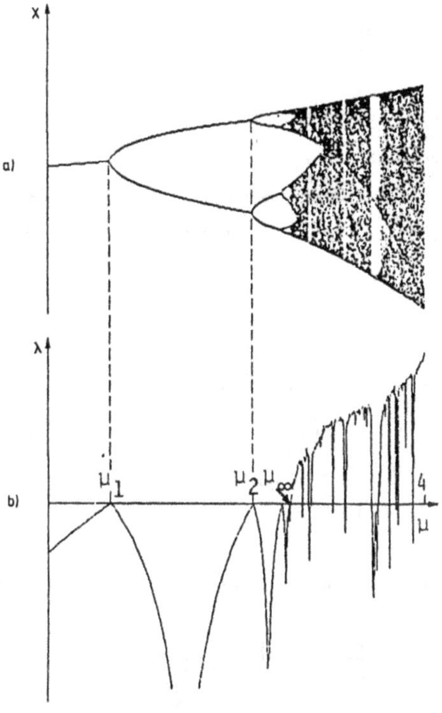

Figure 3.6. (Left) Sequences x_n for 3 values of μ. (a) μ=2.9: period-1 attractor; (b) μ=3.5: period-4 attractor (0.501, 0.875, 0.383, 0.827); (c) μ=3.8: chaotic motion.

Figure 3.7. (Right) (a) The attractor; (b) the Liapunov exponent λ. (reproduced from {Schuster, 1984}, after W. Desnizza).

3.4 SUPERCYCLES, DOUBLING TRANSFORMATION, AND UNIVERSALITY OF THE FEIGENBAUM ROUTE

Supercycles

When

$$\frac{d}{dx} f^{2^n}(x_i^*) = \prod_k f'(x_k^*) = 0, \tag{3.12}$$

the attractor is called maximally stable, superstable, or a supercycle. As the only zero of f' is 1/2, the point x=1/2 belongs to any supercycle, which therefore corresponds to $\mu=M_n$ The distances d_n in Figure 3.5 can thus be written as the differences

$$d_n = f_{M_n}^{2^{n-1}}(\frac{1}{2}) - \frac{1}{2} . \tag{3.13}$$

From now on we shall use a system of coordinates in which the origin is at x=1/2, that is, we change from x to x-1/2. We have then

$$d_n = f_{M_n}^{2^{n-1}}(\frac{1}{2}) . \tag{3.14}$$

From (3.10), there is a finite number D_1 such that

$$\lim_{n\to\infty} (-\alpha)^n d_{n+1} = D_1 ,$$

or, equivalently,

$$\lim_{n\to\infty} (-\alpha)^n f_{M_{n+1}}^{2^n}(0) = D_1 . \tag{3.15}$$

The sequence of iterates $f_{M_n}^{2^n}(0)$, scaled by the factor $(-\alpha)$, converges. This suggests that the scaled functions

$$(-\alpha)^n f_{M_{n+1}}^{2^n}(x/(-\alpha)^n)$$

also converge to a function g_1, that is

$$\lim (-\alpha)^n f^{2^n}_{M_{n+1}} (x/(-\alpha)^n) = g_1(x) \; . \qquad (3.16)$$

Let us now consider this question from a geometrical point of view. In Figure 3.8 we can see the functions f and f^2 at M_1 and f^2 and f^4 at M_2, the lengths d_1 and d_2 being clearly depicted.

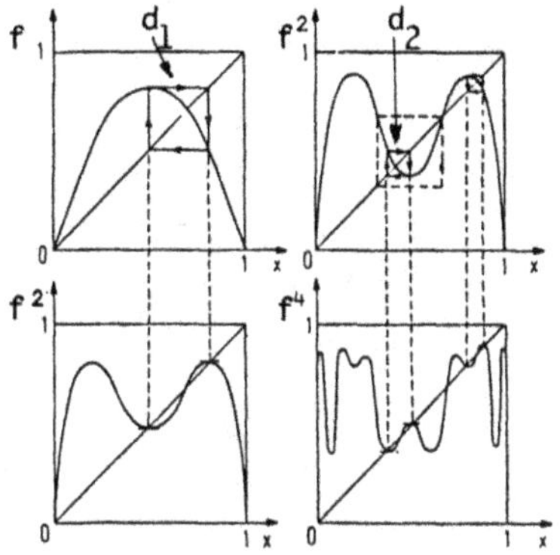

Figure 3.8. Superstable cycles at $\mu=M_1$ and $\mu=M_2$.

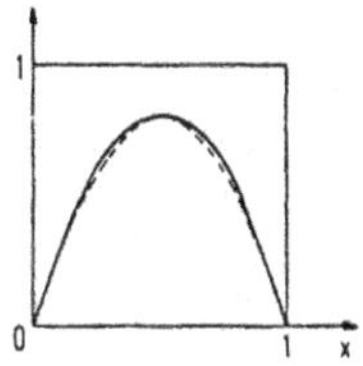

Figure 3.9. Rescaled functions for n=1 and n=2.

(i) The supercycles of period 2 and 4 correspond to squares of sides d_1 and d_2, drawn in full line. In an analogous way d^k is the side of the square which can be drawn in the curve of $f^{2^{k-1}}$ at $\mu=M_k$, with a vertex at the extremum of this function and a diagonal along the bisector y=x.

(ii) Because of (3.10), each of these squares can be obtained by scaling the previous one by the factor $1/(-\alpha)$. Equation (3.16) then means that the restriction of f^{2^n} to these squares converge to a function after rescaling. In Figure (3.9), we can see the first two of these functions.

(iii) In Figure 3.9, it is clear that f^2 in the dashed square depends only on the values of f at distances smaller than d_1 from the maximum. The corresponding result holds also for the functions f^{2^n}. In fact, near the maximum,

$$f^2(x) \simeq f^2(\tfrac{1}{2}) + \tfrac{1}{2} f^{2''}(\tfrac{1}{2}) \; (x-\tfrac{1}{2})^2 \; ,$$

with

$$f^{2}{}''(\tfrac{1}{2}) = f'(\mu/4)\ f''(\tfrac{1}{2}),$$

from which f^2 is determined by the values of f near the maximum and its derivatives at $\mu/4$. Analogously, it turns out that that f^{2^n} is determined by the values of f in an increasingly smaller interval around the maximum and by n derivatives which only fix the absolute scale, not the shape. Their effect is eliminated by the rescaling, this giving an intuitive explanation of the fact that g_1 depends only on the values of f very close to the maximum, in other words of the type of maximum. In particular, it is the same for all the functions with a quadratic one (for the logistic map and for $\lambda \sin \pi x$, for instance). In this sense it is said that g_1 is universal.

(iv) It is convenient to stress that, in the jump from one term to the following one in the sequence (3.16), three different operations are performed:

- forming $f^{2^n} = f^{2^{n-1}} \circ f^{2^{n-1}}$
- shifting μ from M_{n-1} to M_n
- rescaling by the factor $(-\alpha)$.

Doubling transformation

By generalizing (3.16), we can define g_i as

$$g_i(x) = \lim_{n\to\infty} (-\alpha)^n f^{2^n}_{M_{n+i}}\left(\frac{x}{(-\alpha)^n}\right), \quad i = 0,1,\ldots, \tag{3.17}$$

and the doubling transformation T as

$$T\,f(x) = -\alpha\, f(f(x/(-\alpha))). \tag{3.18}$$

The g_i then satify

$$g_{i-1}(x) = T\,g_i(x),$$

because

$$g_{i-1}(x) = \lim_{n\to\infty} (-\alpha)^n f_{M_{n+i-1}}^{2^n} \left(\frac{x}{(-\alpha)^n}\right)$$

$$= \lim (-\alpha)(-\alpha)^{n-1} f_{M_{n-1+i}}^{2^{n-1+1}} \left(-\frac{1}{\alpha}\frac{x}{(-\alpha)^{n-1}}\right)$$

$$= \lim (-\alpha)(-\alpha)^m f_{M_{m+i}}^{2^m} \{\frac{1}{(-\alpha)^m} (-\alpha)^m f_{M_{m+i}}^{2^m} (\frac{1}{\alpha}\frac{x}{(-\alpha)^m})\}$$

$$= -\alpha\, g_i\{g_i(\frac{x}{-\alpha})\}.$$

Let us now take the limit $i\to\infty$:

$$g(x) = \lim_{i\to\infty} g_i(x).$$

Determination of α

It is clear that $g(x)$ is invariant under the doubling operator T, so that

$$g(x) = Tg(x) = -\alpha\, g\{g(-\frac{x}{\alpha})\}, \qquad (3.19)$$

which implies that g does not depend on the map f, but is determined by T. Given a solution $g(x)$ of (3.19), $\lambda g(x/\lambda)$ is another one, which shows that the equation does not fix the scale. If we normalize so that $g(0)=1$, α is universally fixed by

$$\alpha = -\frac{1}{g(1)} . \qquad (3.20)$$

The functional equation (3.19) can be solved numerically. For quadratic maps , a simple estimation of the solution can be obtained by observing that g has the same kind of maximum as f and taking accordingly the approximate form

$$g(x) = 1 + b\,x^2 .$$

By substitution in (3.19), it follows that $b=(-2-\sqrt{12})/4 \simeq -1.366$, $\alpha=|2b|\simeq 2.73$. A more precise calculation was performed by Feigenbaum (1983)[3] with the result

$$g(x) = 1 - 1.52763\, x^2 + 0.104815\, x^4 + \ldots ,$$
$$\alpha = 2.502807876\ldots .$$

The functions g_i represent the highly bifurcated structure of the dynamical system. By studying their limit, we have shown why the constant α is universal for all f with the same type of maximum.

Interpretation of δ

It is clear that g is an unstable fixed point of T. It turns out that δ measures its instability, that is, the speed with which the repeated application of T separates a function from g.

The values $\mu=M_n$ are determined by the condition

$$f_{M_n}^{2^n}(0) = 0 , \tag{3.21}$$

this equation having many solutions, as it also gives the 2^n supercycles in the chaotic region. We can separate those in the periodic regime, by ordering them as $\mu_1 < M_1 < \mu_2 < M_2 < \ldots$. Let us expand f_M around f_{μ_∞}:

$$f_M(x) = f_{\mu_\infty}(x) + (M-\mu_\infty)\, \delta f(x) , \tag{3.22}$$

where

$$\delta f(x) = \frac{\partial f_M(x)}{\partial M}\bigg|_{\mu_\infty} . \tag{3.23}$$

We can then write

$$Tf_M = Tf_{\mu_\infty} + (M-\mu_\infty)\, L_{f_{\mu_\infty}}\, \delta f + O((\delta f)^2), \tag{3.24}$$

where L_f is the linear operator

$$L_f \delta f = -\alpha\, \{f'(f(-x/\alpha))\, \delta f(-x/\alpha) + \delta f(f(-x/\alpha))\}, \tag{3.25}$$

which depends on f. Applying T n times gives

$$T^n f_M = T^n f_{\mu_\infty} + (M-\mu_\infty)\, L_{T^{n-1} f_{\mu_\infty}} \cdots L_{f_{\mu_\infty}}\, \delta f + O((\delta f)^2). \tag{3.26}$$

But as we have seen ,

$$T^n f_{\mu_\infty} \simeq (-\alpha)^n f_{\mu_\infty}^{2^n}\left(\frac{x}{(-\alpha)^n}\right) = g(x), \quad n \gg 1 , \tag{3.27}$$

from which (3.26) can be written approximately

$$T^n f_M = g(x) + (M-\mu_\infty)\, L_g^n\, \delta f, \quad n \gg 1. \tag{3.28}$$

Now, it turns out that L_g has the property that only one of its eigenvalues δ_k, say $\delta_1 = \delta$, has a modulus larger than unity, so that $|\delta_i| < 1$ for $i \neq 1$. This implies that only the corresponding eigenfunction u_1 is relevant in (3.28) and that if $\delta f = \Sigma\, a_k u_k$, then

$$T^n f_{M_n}(x) = g(x) + (M_n - \mu_\infty)\, \delta^n a_1 u_1(x), \quad n \gg 1. \tag{3.29}$$

Taking x=0 ,

$$T^n f_{M_n}(0) = g(0) + (M_n - \mu_\infty)\, \delta^n a_1 u_1(0), \tag{3.30}$$

and recalling from (3.21) that

$$T^n f_{M_n}(0) = (-\alpha)^n f_{M_n}^{2^n}(0) = 0$$

and that g(0)=1, it follows that

$$(M_n - \mu_\infty)\ \delta^n = \frac{-1}{a_1 u_1(0)} = \text{const}, \tag{3.31}$$

which is precisely (3.9'). To obtain (3.9), we introduce the slopes

$$\nu = \frac{d}{dx_1^*} f_\mu^{2^n}(x_1^*) = \Pi\ f'(x_k^*). \tag{3.32}$$

Given n and ν , the value of μ is determined and conversely. We may thus parametrize the problem with the pair (n, ν). From (3.29) we have then

$$g_{o,\nu}(x) = g(x) + (M_{n,\nu} - \mu_\infty)\ \delta^n\ a_1 u_1(x), \tag{3.33}$$

where $g_{0,\nu}$ is again a universal function:

$$g_{o,\nu}(x) = \lim_{n\to\infty}\ (-\alpha)^n\ f_{M_{n,\nu}}^{2^n}\ (\frac{x}{(-\alpha)}n). \tag{3.34}$$

As $\mu_o = M_{n,1}$, since the slope is one at the bifurcation, it follows that

$$M_n - \mu_\infty = \text{constant}\ \delta^{-n}\ , \quad n >> 1,$$

which is (3.9).

We have thus given arguments which show that α and δ are constants which depend only on the kind of maximum of f. To complete the proof, it is necessary to study in detail Equation (3.19) and to demonstrate that δ is indeed the only eigenvalue of L_g with modulus greater than one. The reader is referred for these points and for more complete discussions to the works of Feigenbaum (1978, 1983)[3),4)], Collet and Eckmann (1980)[1)] and Lanford (1980)[5)].

Conservative Feigenbaum sequences

The logistic map is not conservative. Curiously enough, it has been shown that in the case of conservative systems, infinite sequences of period doublings, similar to the previous ones, also appear.[6] However, the α and δ numbers take the values $\alpha = 4.018076\ldots$, $\delta = 8.721097\ldots$, different from those of the nonconservative case.

3.5 EXPERIMENTAL EVIDENCE OF THE FEIGENBAUM ROUTE

The subharmonic cascade appears in many processes which can be observed in the laboratory as well as in many theoretical models, in very different physical contexts. As we have seen in the previous exposition, its most characteristic properties are the following:

(i) When a parameter changes, there is an infinite sequence of period doublings, in which subharmonics of frequency $f_0/2^n$ appear, f_0 being the fundamental frequency.

(ii) The control parameter scales as

$$\mu_n - \mu_\infty = \text{cons.}\ \delta^{-n}.$$

(iii) There is a unidimensional Poincaré map with a single maximum.

Although they have not been discussed, it is worth mentioning two more properties.

(iv) It is clear that the intensity of each new odd subharmonic must be smaller than that of the preceding one. It can be proved that their ratio is given approximately by the factor 0.1525, which corresponds to 8.17 decibels (as $-10\log_{10}(0.1525) = 8.17$) (Collet and Eckmann, 1980,[1]) {**Schuster, 1984**}).

(v) External noise destroys the fine structure of the power spectrum, but as a consequence of (iv) one more subharmonic can be observed by reducing the noise level by a factor 0.1525.

In the following we will consider some systems which manifest this phenomenon.

(a) Navier-Stokes equation in a torus[7)]

These authors considered the Navier-Stokes equation for an incompressible fluid moving in a bidimensional torus:

$$\frac{\partial \vec{u}}{\partial t} + (\vec{u}.\vec{\nabla})\,\vec{u} = -\,\vec{\nabla}p + \vec{f} + \nu\,\Delta\vec{u}, \qquad \vec{\nabla}.\vec{u} = 0, \tag{3.35}$$

where $\vec{u}$ is the velocity, $\vec{f}$ the volume force, p the pressure and ν the viscosity. They performed a Fourier expansion :

$$\vec{u} = \sum e^{i\vec{k}.\vec{x}}\; x_{\vec{k}}\,\frac{k^{\perp}}{|\vec{k}|}\,, \qquad k^{\perp} = (k_2,k_1), \tag{3.36}$$

$$\vec{x}_k = -\vec{x}_{-k}$$

and kept only the five modes k = (1,1), (3,0), (2,-1), (1,2), (0,1) . They found that the corresponding amplitudes x_k satisfy

$$\begin{aligned}
\dot{x}_1 &= -2x_1 + 4x_2x_3 + 4x_4x_5\,,\\
\dot{x}_2 &= -9x_2 + 3x_1x_3\,,\\
\dot{x}_3 &= -5x_3 - 7x_1x_2 + r\,,\\
\dot{x}_4 &= -5x_4 - x_1x_5\,,\\
\dot{x}_5 &= -x_5 - 3x_1x_4\,,
\end{aligned} \tag{3.37}$$

r being the control parameter which depends on $\vec{f}$. Equation (3.37) was solved numerically with the following results:

Ar r=22.8537 a period T solution C_0 appears. At higher values of r other solutions C_1, C_2, C_3 with periods 2T, 4T, 8T are observed. Another cascade C^*_1, i=0, ,4, with periods T*, 2T*, 4T*, 8T*, 16T* starts at r=28.6662. After r≃28.9 the motion is chaotic.

Defining $\delta_n = (r_n - r_{n-1})/(r_{n-1} - r_n)$, the sequences for both cascades are

δ_1	δ_1^*
24.22	2.57
9.47	4.63
6.54	4.64
5.29	4.42
4.73	

The authors warn that the last value in the second column is not completely reliable because of the very long period of the solution. As we see, the tendency to δ is clear.

The analogous sequences of α values are

α_1	α^*_1
4.51	2.34
3.39	2.63
2.91	2.44
2.53	

It seems clear that the Feigenbaum mechanism is operating here. However, it must be stressed that the precise way in which this occurs is not well understood.

(b) The Duffing oscillator.

The Duffing equation is

$$\ddot{x} + k\dot{x} + x^3 = b \sin 2\pi t . \tag{3.39}$$

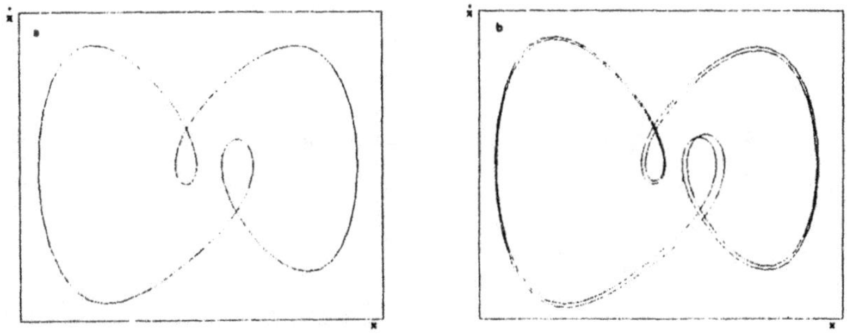

Figure 3.10. Attractors of period 1 and 2 in the Duffing oscillator (after Feigenbaum, 1983 [3]).

In Figure 3.10, we can see a period-1 attractor (the time unit is the period of the external force). When the parameter k is increased, period doublings are produced at $k_1, k_2, \ldots, k_n, \ldots$. The sequence $\delta_n = (k_n - k_{n-1})/(k_{n+1} - k_n)$ seems to converge to δ, δ_3 being already equal to 4.69... .

(c) The Rayleigh-Bénard Convection

Libchaber and Maurer [8] studied in 1980 a Bénard cell with two convection rolls in liquid helium. Their results appear in Figure 3.11. When the temperature difference is increased, peaks of frequencies f, f/2, f/4, f/8, and f/16 appear in the power spectrum. Moreover, the intensity of consecutive subharmonics differ by a value close to 10 dB (remember the theoretical prediction: 8.2 dB).

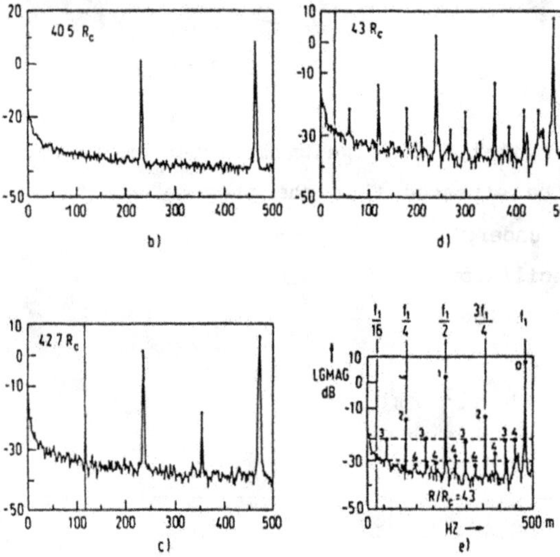

Figure 3.11. Bénard experiment by Libchaber and Maurer.[8] Shown are the power spectra for several increasing values of the Rayleigh number.

(d) Nonlinear RCL circuit

Linsay [9] considered a nonlinear RCL oscillator, as represented in Figure 3.11, the nonlinearity being provided by the capacitor diode which has a capacitance varying as

$$C(V) = C_0/(1 + V/\phi)^{\gamma} , \tag{3.40}$$

with C_0=81.8 pF, ϕ=0.6 V and γ= 0.44. The differential equation of the circuit is

$$L\ddot{q} + R\dot{q} + V(q) = V_0 \sin (2\pi ft). \tag{3.41}$$

Linsay obtained the following results: (i) if $I = \dot{q}$ and $I_n = I(t_0 + nT)$, $T=1/f$, the function $g(x)$ such that $I_{n+1} = g(I_n)$ has a quadratic maximum; (ii) there is a cascade of period doublings for increasing values of V_0, as can be observed in the power spectrum. If the values of δ_n are calculated, δ_1 and δ_2 differ from δ by less than 5%; (iii) the intensities of the subharmonics coincide with the prediction of Feigenbaum's theory.

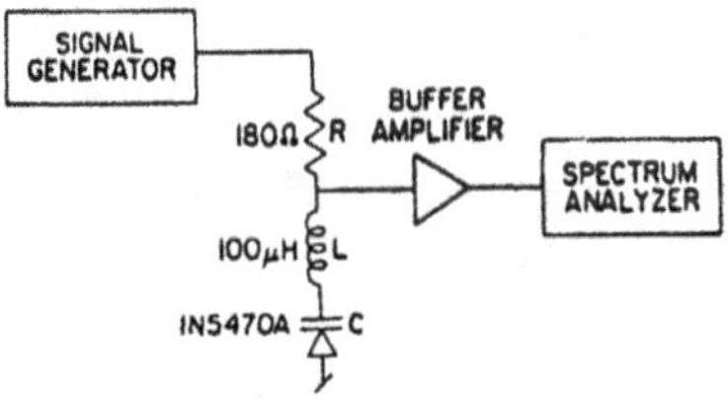

Measured value of the convergence rate.

Subharmonic	ΔV	δ_n
f/2	3.2	
f/4	0.72	4.4
f/8	0.16	4.5

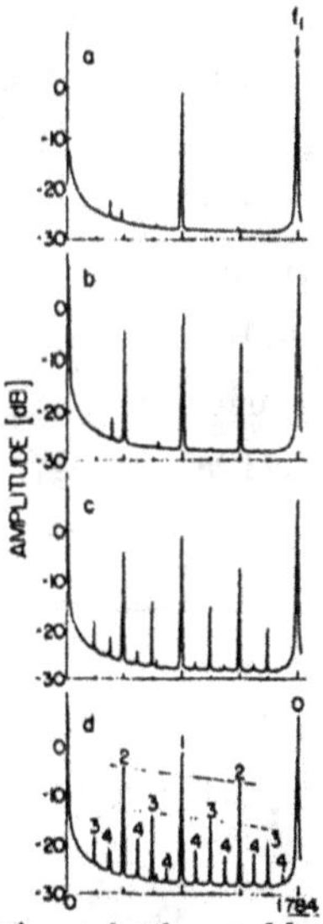

Figure 3.13. Nonlinear RCL oscillator (left) and the subharmonic spectrum for succesive period doublings; the final one is compared with theory (after Linsay, 1981[9]).

CHAPTER 4 INTERMITTENCIES

4.1 DESCRIPTION OF THE PHENOMENON

There is another route to chaos, usually called the Manneville-Pomeau route or scenario, which proceeds through the appearance of intermittencies.

When a signal alternates randomly between long regular or laminar phases (the so called intermissions) and relatively short irregular bursts, it is said that the motion is intermittent or that there is an intermittency. In general, when a control parameter changes, the number of chaotic bursts varies. To be more precise, below a certain threshold the motion is regular, but as the parameter separates from this critical value, chaotic bursts appear, the length of the regular phases decreasing, so that at the end the regular motion disappears.

This intermittent behaviour was first found by Pomeau and Manneville[1)] in 1979 in the Lorenz model, as is shown in Figure 4.1, where the evolution of the variable Y is plotted for several increasing values of r. For $r<r_o$ =166.07, the motion is periodic. Above this critical value, the regular oscillations are interrupted by erratic bursts with a frequency which increases with r, until a truly chaotic motion is achieved. Intermitencies appear also in the logistic map

$$x_{n+1} = \mu x_n (1 - x_n).$$

It has been found that a period-3 window opens at $\mu_o = 1 + \sqrt{8}$, as is shown schematically in Figure 4.2. In Figure 4.3, we can see the iterates of the logistic map near this critical value. In (a), $\mu = \mu_c+0.002$ while in (b) $\mu = \mu_c-0.002$. It is clear that in the first case the behaviour is regular, while in the second one it is intermittent. In (c), the first 200 third iterates of a generic x are plotted. The regular phases are the horizontal lines.

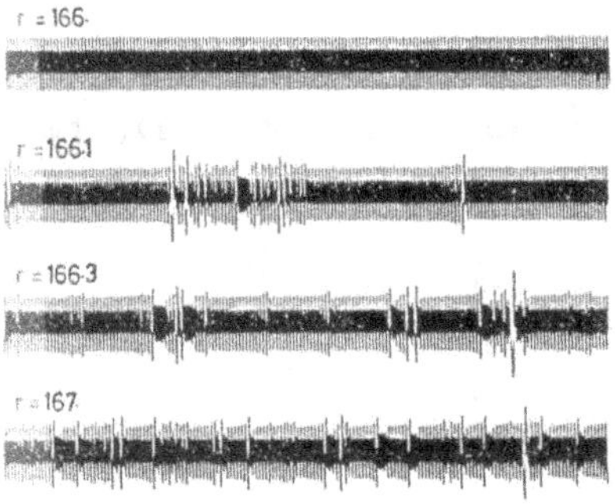

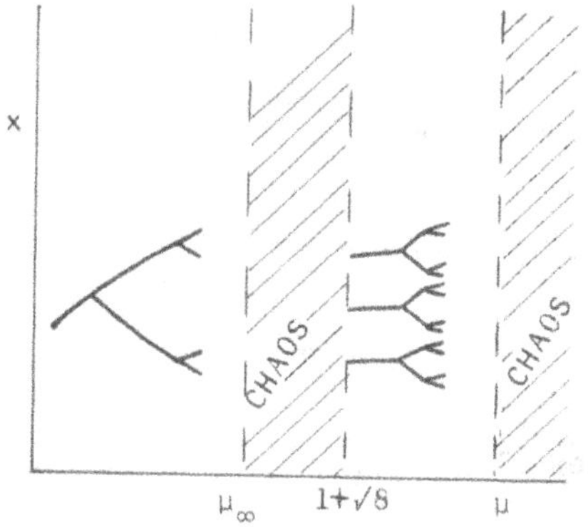

Figure 4.1. Intermittency in the Lorenz model (after Pomeau and Manneville, 1980[2]) (left).

Figure 4.2. Period-3 window in the logistic map (schematically) (right).

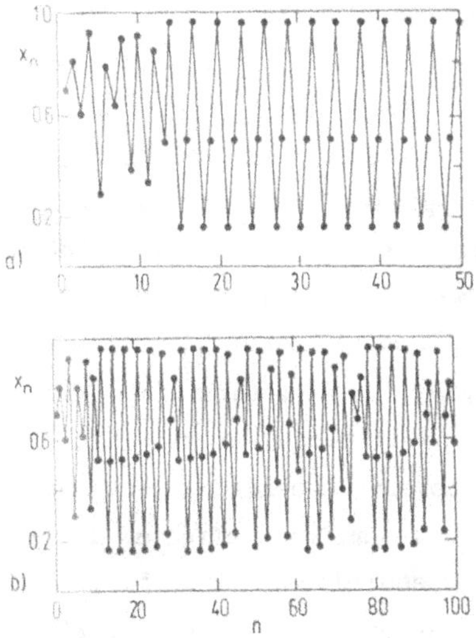

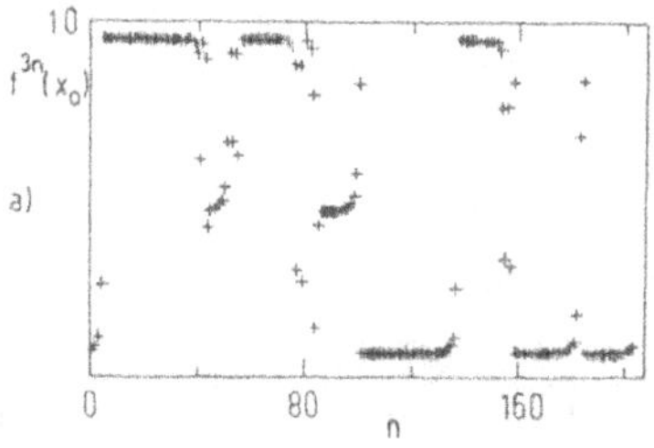

Figure 4.3. Iterates of the logistic map near $\mu_0 = 1 + \sqrt{8}$ (after Hirsh, Huberman and Scalapino, 1981 [4] see text).

Pomeau and Manneville gave in 1980[2] the following interpretation of the phenomenon: below the critical value of the parameter, the dynamical system describes a limit cycle and performs regular oscillations which are stable under small perturbations. The behaviour of the system can then be explained by Floquet theory, [3] according to which the linearized variational equations around the limit cycle take the form

$$\dot{x} = A(t)\,x, \tag{4.1}$$

where the matrix $A(t)$ satisfies $A(t+T)=A(t)$, T being the period of the limit cycle. If $B(t)$ is a fundamental matrix of (4.1), that is one having as columns n independent solutions, it can be shown that

$$B(t) = Q(t)\exp(tR)\,, \tag{4.2}$$

where $Q(t+T)=Q(t)$ and R is a constant matrix. It follows that

$$B(t+T) = B(t)\,C\,, \tag{4.3}$$

where $C=\exp(tR)$ is referred to as the Floquet or monodromy matrix of $B(t)$. Its eigenvalues λ_k are called the characteristic or Floquet multipliers and determine the stability properties of the unperturbed solution. The eigenvalues ρ_k of R, are the characteristic exponents and verify $\lambda_k=\exp(T\rho_k)$.

As long as the multipliers are inside the unit circle in the complex plane, the periodic solution is stable. But, if the control parameter changes, they move in this plane and, when one of them crosses the circle, a bifurcation is produced and the solution becomes unstable. There are three different ways in which this can happen, according to whether the circle is crossed through +1, two complex conjugate values, or -1. Accordingly, the intermittencies are classified as being of type I (+1), type II ($\alpha\pm i\beta$), or type III (-1). It is necessary to stress that the phenomenon depends not only on the linearized equation, but also on the weak nonlinear effects. In fact, an intermittency can only arise when these effects tend to increase the instability, a case referred to sometimes as subcritical bifurcation. This depends of the sign of the leading nonlinear terms, as we will see. For instance, in the case of the period doubling, one has $f^{2\prime}=+1$, $f^{2\prime\prime}=0$, $f^{2\prime\prime\prime}<0$.

4.2 THEORY OF TYPE I INTERMITTENCY

In this type, one of the Floquet multipliers crosses the unit circle through +1. In order to make a theory which accounts for this, two points must be separately discussed: the local behavior and the mechanism for the reinjection after the chaotic bursts.

Local Analysis

In the generic case, the eigenvalue which goes across the circle is simple. If we take the coordinate x along its eigendirection, the relevant part of the dynamical system can be written as

$$x_{n+1} = f_\lambda(x_n) .$$

Type I intermittencies correspond to the so-called tangent bifurcations in which a stable and an ustable fixed point coalesce and disappear when the function f(x) is tangent to the bisector y=x, as is depicted in Figure 4.4.

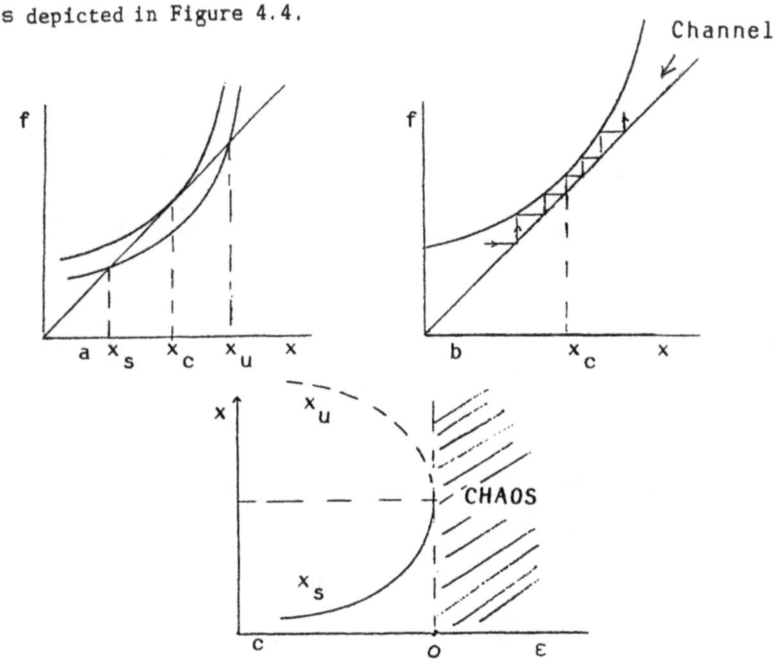

Figure 4.4. Type I intermittency (see text).

Near the point at which f'=+1 and the critical value of the control parameter λ, the map f can be written in generic form as

$$u_1 = u + u^2 + \epsilon \tag{4.5}$$

with $u = x - x_c$, after a suitable change of scale. We will impose the sole restriction that $f'' > 0$, thus discarding the case of period doubling. The fixed points are the solutions of $u_1 = u$, that is, $u = \pm\sqrt{-\epsilon}$, and they only exist if $\epsilon < 0$. As $du_1/du = 1+2u$, $u = +\sqrt{-\epsilon}$ is unstable, and $u = -\sqrt{-\epsilon}$ is stable.

Now, if $\epsilon < 0$, there is a stable point u_-. If ϵ becomes positive it disappears, but there remains a kind of memory of it in the form of a channel between f(x) and the bisector y=x. As can be seen in Figure 4.4 (b), it may take many iterations before x_n gets out of the channel and during this process its value is nearly constant, which results in an almost regular oscillation. In Figures 4.5 and 4.6, we can see the situation for the logistic map and the Lorenz model. It is clear that the smaller the positive value of ϵ, the longer the laminar phase.

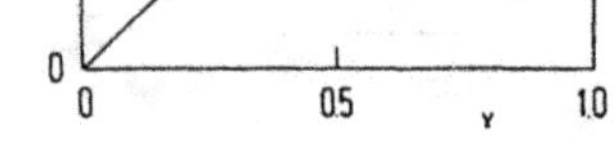

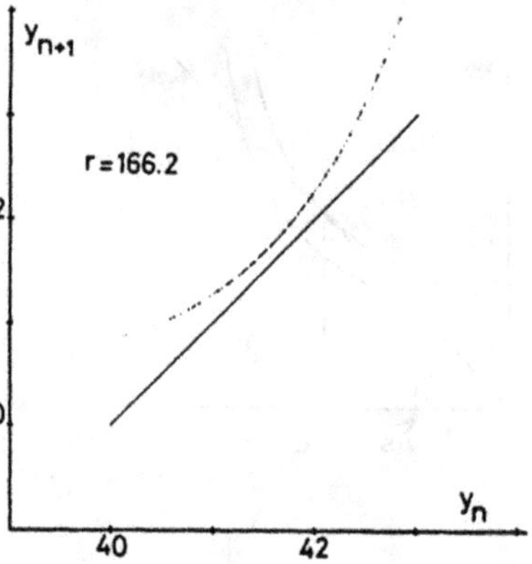

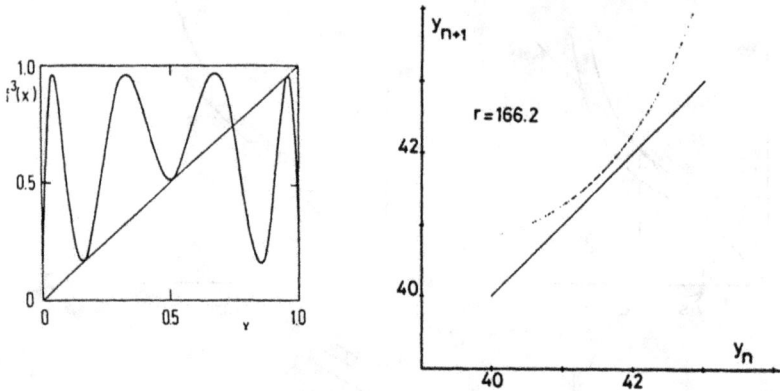

Figure 4.5. Third iterate of the logistic map for $\mu = 1+\sqrt{8}$ (left).
Figure 4.6. Poincaré map of the Lorenz model for r slightly above $r_c = 166.02$ (after Pomeau and Manneville, 1980[2)]) (right).

Length of the Laminar Phases

In order to accept the previous model as a precise account of the phenomenon, it is necessary to show that it leads to predictions which agree with experiment. Indeed, there are such predictions: the average length of the laminar phases and the distribution of these lengths.

We have stated that they become longer when the parameter goes to zero, but what is the precise form of this dependence? We will answer this question by making a simplification. If u_1 is close to u in (4.5), we can interpret the difference $u_1 - u$ as the derivative du/dk, where the index of the iteration k is taken as a continuous variable. This is permissible as long as u changes slowly so that there are many iterations in the phase. This amounts to considering, instead of (4.5), the ordinary differential equation

$$\frac{du}{dk} = \epsilon + u^2 , \qquad (4.6)$$

the general solution of which is

$$u(k) = \sqrt{\epsilon} \tan(\sqrt{\epsilon}\ (k-k_0) - \arctan(u_0/\sqrt{\epsilon})). \qquad (4.7)$$

Let us take, for simplicity, $k_0 = 0$. The function u(k) diverges for

$$k = (\pi/2 + \arctan(u_0/\sqrt{\epsilon}))/\sqrt{\epsilon} .$$

This is no longer valid when u is not small; however it us with a scaling law: the average length of the laminar phases diverges as

$$\epsilon^{-1/2} \sim |\mu - \mu_c|^{-1/2} .$$

To be more precise, let c be the maximum value of u in the laminar regime. The length L of the phase is then

$$L = \frac{1}{\sqrt{\epsilon}} \ (\arctan \frac{c}{\sqrt{\epsilon}} - \arctan \frac{x_0}{\sqrt{\epsilon}}) .$$

The probability distribution P(L) depends on the probability of reinjection at x_0, $P^*(x_0)$, through provides

$$P(L) = P^*(x_0)\ |dx_0/dL|$$

By taking a constant distribution $P^*=1/2c$, it follows that

$$P(L) = (\epsilon/2c)\ (\ 1 + \tan^2(\arctan\ (c/\sqrt{\epsilon})-\sqrt{\epsilon}l)),\ L \leq L_{max},$$

and

$$\langle\ L\ \rangle = \int_0^{L_{max}} dL\ P(L)\ L \sim \epsilon^{-1/2} \ \text{for} \ \ \epsilon\to 0\ .$$

Although $P^*(x_0)$ may depend on the particular model, $\langle L \rangle$ is always of order ϵ .

Reinjection

Once the system leaves the channel, the approximation made so far is no longer valid. For this reason, it is difficult to explain how the signal comes back to the mouth of the channel, a process known as reinjection or relaminarization. As the system explores finite regions of the phase space, it is necessary to resort to topological arguments to explain what happens. We will describe here the process in two particular cases.

(a) Flow in the torus T^2: The Poincaré map of a flow in a torus is noninvertible map of the circle on itself. It is represented in Figure 4.7, where the inverse tangent bifurcation leading to intermittent behaviour is depicted. It is clear that the mechanism of reinjection consists in the successive iterations giving a turn to the circle. But two remarks are in order. First, this intermittent behavior cannot give a transition to turbulence, since the flow on a bidimensional torus is always regular according to the Poincaré-Bendixson theorem. The bursts appear thus in an almost regular way, the motion being either periodic or quasiperiodic. Second, when the channel closes upon itself, there is a stable fixed point. Consequently, there is frequency locking.

(b) Twice covering map of the unit interval on itself: An example of this kind could be the Bernouilli shift $\sigma(x)=2x(\mathrm{mod}\ 1)$. The general case is represented in Figure 4.8. As $f(x)$ runs twice along the unit interval, it has necessarily a fixed point which has been taken as the origin. In this case the trajectory explores the unstable regions of phase space ($|f'(x)| > 1$).

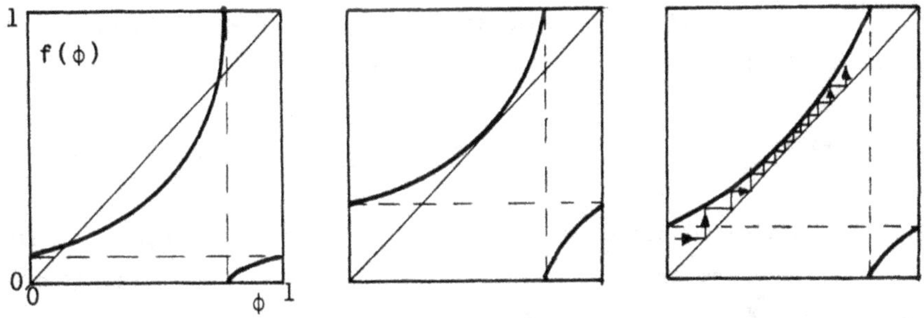

Figure 4.7. Type I intermittency in the return map of a flow on a torus T^2.

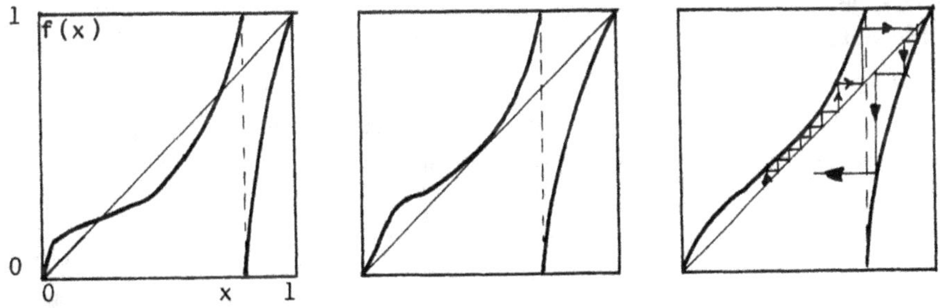

Figure 4.8. Reinjection after a type I intermittency in a twice covering map on the unit interval.

4.3 THEORY OF TYPE III INTERMITTENCY

The main features of the intermittent motion are almost independent of the type. In the case of type III, the relevant eigenvalue crosses the unit circle through -1. Accordingly, we can express the map f(x) near the fixed point as

$$x \to f(x) = -(1+\epsilon)x + \alpha x^2 + \beta x^3 \quad , \tag{4.8}$$

up to first order in ϵ and to third order in x(it is enough to consider the ϵ-dependence of the terms which are linear in x). This kind of approximation in which one keeps some nonlinearity is sometimes called *à la Landau*. The situation seems analogous to the period doubling one studied in Chapter 2. Let us consider the second iterate:

$$f^2(x) = f(f(x)) = (1 + 2\epsilon)x + \beta' x^3, \quad \beta' = -2(\beta+\alpha^2), \tag{4.9}$$

where the quadratic terms in ϵ have been neglected as well as others in ϵx^2 and ϵx^3. We see that the coefficient of the linear term is now close to +1. Now, it turns out that the effect of the cubic term in (4.9) depends strongly on the sign of $\beta' = f^{2'''}(0)/6$. If $\beta'<0$, this term has opposite sign to the linear one, as for the pitchfork bifurcation with period doubling, and the bifurcation is then called supercritical. This case has been considered in Chapter 2. On the other hand, if $\beta'>0$, the nonlinearity strengthens the instability, the bifurcation being subcritical and corresponds to intermittent behavior.

As in the case of type I intermittencies, we can approximate $x''-x=f^2(x)-x$ by dx/dk and write (4.9) as

$$dx/dk = x(2\epsilon + \beta' x^2), \tag{4.10}$$

the general solution of which is

$$k = \frac{1}{2\epsilon}\{\log \frac{\beta' x^2}{\beta' x^2+2\epsilon} - \log \frac{\beta' x_0^2}{\beta' x_0^2+2\epsilon}\} . \tag{4.11}$$

Proceeding as in the case of type I, it is found that

$$P(L) \sim \epsilon^{3/2}\exp(4\epsilon\bar{L})/(\exp(4\epsilon\bar{L}) - 1)^{3/2}, \quad \langle L \rangle \sim \epsilon^{-1}. \tag{4.12}$$

4.4 TYPE II INTERMITTENCIES

In this case, a pair of complex conjugate eigenvalues crosses the unit circle. Although the phenomenology is very similar to that of the other types, the analysis is more involved, since two variables instead of one must be used. We will not discuss this type of intermittency here, but will refer to (Bergé el al., 1984).

4.5 EXPERIMENTAL OBSERVATION OF INTERMITTENCIES

Intermittencies have been found in many different experiments. We can see in Figures 4.9 and 4.10 two examples, both referring to the Rayleigh-Bénard convection. Figure 4.9 depicts the variation of the vertical component of the velocity in the middle of a cell, for increasing values of the Rayleigh number (after Bergé et al., 1980).[5] It is an intermittency of type I. Figure 4.10 shows results of measurements of the temperature gradient (after Dubois et al., 1983 [6]). This intermittency is of type III.

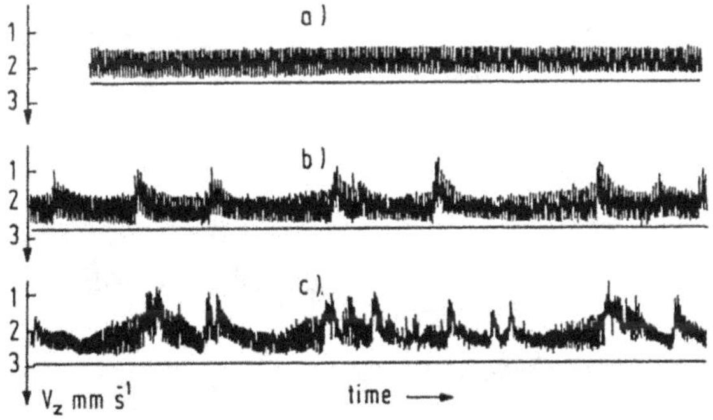

Figure 4.9. Intermittency of type I (see text).

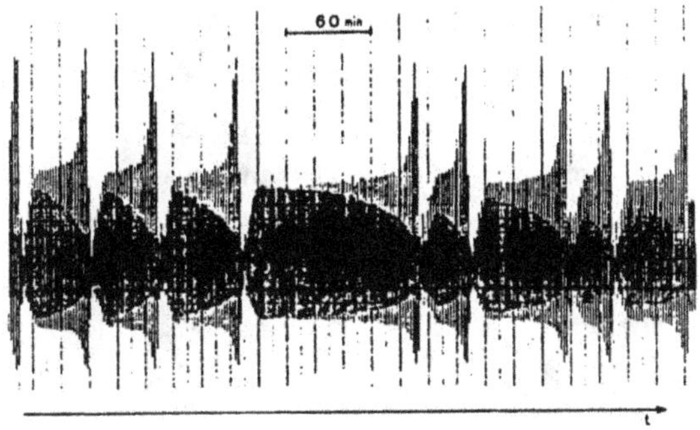

Figure 4.10. Intermittency of type III (see text).

REFERENCES

REFERENCES OF GENERAL CHARACTER*

Arnold, V.I., and Avez, A., Ergodic Problems of Classical Mechanics, (Benjamin, New York, 1968).

Bergé, P., Pomeau, Y., and Vidal, Ch., L'ordre dans le chaos (Hermann, Paris, 1984).

Berry, M., edited by S. Jorna. Topics in Nonlinear Dynamics, A.I.P. Conf. Proc. 46 (1978).

Haken, H., Advanced Synergetics (Springer, Heidelberg-New York, 1984).

Helleman, R.H.G., "Self Generated Chaotic Behavior in Nonlinear Mechanics", in Fundamental Problems in Statistical Mechanics, Vol. 5, edited by E.G.D. Cohen (North-Holland, Amsterdam, 1980), p. 165.

Chaotic Behavior in Deterministic Systems, edited by R.H.G. Helleman and G. Iooss (North-Holland, Amsterdam, 1983).

Lichtenberg, A.J., and Liebermann, M.A., Regular and Stochastic Motion, (Springer, Heidelberg-New York, 1983).

Schuster, H.G., Deterministic Chaos: An Introduction (Physik-Verlag, Weinheim, 1984).

CHAPTER 1.

[1] Poincaré, H., Les méthodes nouvelles de la mécanique céleste (Gauthiers-Villars, Paris, 1892).

[2] Einstein, A., Verh. Deut. Phys. Ges. 19, 82 (1917).

[3] Kolmogorov, A.N., "On Conservation of Conditionally-Periodic Motions Under Small Perturbations of the Hamiltonian", Dokl. Akad. Nauk, SSSR 98, 527 (1954).

[4] Arnold, V.I., "Small Denominators II, Proof of a Theorem of N. Kolmogorov", Russ. Math. Surveys 18, 5 (1963).

[5] Moser, J., "Convergent Series Expansions of Quasi-Periodic Motions", Math. Ann. 169, 163 (1967).

---

*These will be enclosed in curly brackets, e.g., {Berry, 1978}.

[6] Lorenz, E.N., "Deterministic Nonperiodic Flow", J. Atmos, Sci, 20, 130 (1963).

[7] Huberman, B.A., Crutchfield, J.P., and Packard, N.H., "Noise Phenomena in Josephson Junctions", Appl. Phys. Lett. 37, 750 (1980)

[8] D'Humières, D., Beasly, M.R., Huberman, B.A., and Libchaber, A., "Chaotic States and Routes to Chaos in the Forced Pendulum", Phys. Rev. 26A, 3483 (1982).

[9] Sitnikov, K.A., "Existence of Oscillating Motions for the Three-Body Problem", Dokl, Akad. Nauk. 133, 303 (1960).

[10] Siegel, C.L., and Moser, J., Lectures on Celestial Mechanics (Springer, Berlin, 1971).

[11] Eminhizer, C.R., cited in (Helleman, 1980), p.177.

[12] Epstein, I.R., Dustin, K., Depper, P. de, and Orban, M., "Oscillating Chemical Reactions", Sci. Am. 248, 112 (1983).

[13] Roux, J.C. Rossi, A., Bachelart, S., and Vidal, Ch., "Experimental Observation of Complex Behavior During a Chemical Reaction", Physica 2D, 395 (1981).

[14] Ford, J., "How Random is a Coin Toss?", Physics Today 36, 40 (1983).

[15] Siegel, C.L., "On the Integrals of Canonical Systems", Ann. Math. 42, 806 (1941); Siegel, C.L., "Über die Existenz einer Normalform analytischer Hamiltonischer Differentialgleichungen in der Nähe einer Gleichgewichtslösung", Math. Ann. 128, 144 (1954).

[16] Moser, J., Lectures on Hamiltonian Systems, Memoirs Am. Math. Soc. 81, 1 (1968).

[17] Moser, J. (1973), Stable and Random Motions in Dynamical Systems (Princeton University Press, Princeton, 1973).

[18] Swinney, H.L. and Gollub, J.P., "The Transition to Turbulence", Phys. Today 31, 41 (1978).

[19] Hénon, M., and Heiles, C., "The Applicability off the Third Integral of Motion: Some Numerical Experiments", Astron. J. 69, 73 (1964).

[20] Gustavson, F., "On Constructing Formal Integrals of a Hamiltonian System Near An Equilibrium Point", Astron, J. 21, 670 (1966).

[21] Shannon, C.E., and Weaver, W. (1949), The Mathematical Theory of Information (University of Illinois Press, Urbana, 1949).

[22] López, A., "Trayectorias Caóticas de Osciladores", Complutense University, Madrid (1985). (unpublished).

[23] Pesin, Ya. B., "Characteristic Lyapunov Exponents and Smooth Ergodic Theory', Russ. Math. Surveys 32(4), 55 (1977).

[24] Farmer, J.D., "Information Dimension and the Probabilistic Structure of Chaos", Z. Naturforsch. 37a, 1304 (1982).

CHAPTER 2.

[1] B. Mandelbrot (198, The Fractal Geometry of Nature (Freeman, New York, 1983).

[2] Farmer, J.D., Ott. E., and Yorke, J.A., "the Dimension of Chaotic Attractors", Physica 7D, 153 (1983).

[3] Ruelle, D., and Takens, F., "On the Nature of Turbulence", Commun. Math. Phys. 20, 167 (1971).

[4] Ruelle, D., "Strange Attractors", Math. Intelligencer 2, 126 (1980); "Les attracteurs étranges", La recherche 108, 132 (1980).

[5] Hénon, M., "A Two-Dimensional Map With a Strange Attractor", Commun. Math. Phys. 50, 69 (1976).

[6] Lorenz, E.N., "Deterministic Nonperiodic Flow", J. Atmos. Sci. 20, 130 (1963).

[7] Lanford, O.E., "Computer Pictures of the Lorenz Attractor", Appendix to Williams, R.F., "The Structure of Lorenz Attractors", in Turbulence Seminar Berkeley 1976/1977, Lecture Notes in Mathematics 615, edited by P. Bernard and T. Ratiu (Springer, New York, 1977), p. 94.

[8] Grassberger, P., and Procaccia, I., "On the Characterization of Strange Attractors", Phys. Rev. Lett. 50, 346 (1983).

[9] Grassberger, P., and Procaccia, I., "Measuring the Strangeness of a Strange Attractor", Physica 9D, 189 (1983).

[10] Hentschel, H.G., and Procaccia, I., "The Infinite Number of Dimensions of Strange Attractors", Physica 8D, 435 (1983).

[11] Takens, F., "Detecting strange Attractors in Turbulence", in Dynamical systems and Turbulence, Warwick 1980, Lecture Notes in Mathematics 898 (Springer, Berlin, 1981), p. 366.

[12] Kaplan, J.L., and Yorke, J.A., "Chaotic Behavior of Multidimensional Difference Equations", in Functional Differential Equations and Approximation of Fixed Points, edited by H.O. Peitgen and H.O. Walther, Lectures Notes in Mathematics 730 (Springer, New York, 1979), p. 2280.

[13] Landau, L.D., D. R. Dokl. Acad. Sci. USSR 44, 311 (1944).

[14] Landau, L.D. and Lifshitz, E.M., Fluid Mechanics (Pergamon, Oxford. 1959).

[15] Newhouse, S., Ruelle, D., and Takens, F., "Occurrence of Strange Axiom-A Attractors Near Quasi-Periodic flows on $T_m, m \leq 3$", Commun. Math. Phys. 64 ,35 (1978).

[16] Eckman, J.P., "Roads to Turbulence in Dissipative Dynamical Systems", Rev. Mod. Phys. 53, 643 (1981).

[17] Ott, E., "Strange Attractors and Chaotic Motions of Dynamical Systems", Rev. Mod. Phys. 53, 655 (1981).

[18] Swinney, H.L., and Golub, J.P., "The Transition to Turbulence", Phys. Today 31(8), 41 (1978).

[19] Grebogi, C., Ott, E., and Yorke, J., "Are Three-Frequency Quasi-Periodic Orbits to be Expected in Typical Nonlinear Systems?", Phys. Rev. Lett. 51, 339 (1983).

[20] Libchaber, A., Fauve, S., and Laroche, C., "Two-Parameter Study of the Routes to Chaos", Physica 7D, 73 (1983).

[21] Martin, S., Leber, H., and Martienssen, W., "Oscillations and Chaotic States of the Electrical Conduction in BSN Crystals", Phys. Rev. Lett. 52, 303 (1984).

CHAPTER 3.

[1] Collet, P., and Eckmann, J.P., Iterated Maps of the Interval as Dynamical Systems (Birkhäuser, Boston, 1980).

[2] Verhulst , P.F., "Recherches mathématiques sur la loi d'accroisement de la population" Mém. Acad. Roy. Bruxelles 20,1 (1845).

[3] Feigenbaum, M.J., "Universal Behavior in Nonlinear Systems", Physica 7D, 16 (1983).

[4] Feigenbaum, M.J., "Quantitative Universality for a Class of Nonlinear Transformations", J. Stat. Phys. 19, 25 (1978); "The Onset Spectrum of Turbulence", Phys. Lett. 74A, 375 (1979) "The Transition to Aperiodic Behavior in Turbulent Systems", Commun. Math. Phys. 77, 65 (1980).

[5] Lanford, O.E., "A Computer Assisted Proof of the Feigenbaum Conjecture", Bull. Am. Math. Soc. 6, 427 (1982).

[6] Hellemann, R.H.G., in Long Term Predictions in Dynamics, edited by C.W. Horton, J.R. Reichl, and A.G. Szebehely (Wiley, New York, 1983).

[7] Boldrighini, C., and Franceschini, V., "A Five-Dimensional Truncation of the Navier-Stokes Equations", Commun. Math. Phys. 64, 159 (1979).

[8] Libchaber, A., and Maurer, J., "Une expérience de Rayleigh-Bénard de géometrie réduite", J. Phys. (Paris) Coll. 41C, 3 (1980).

[9] Linsay, P.S., "Period Doubling and Chaotic Behavior in a Driven Anharmonic Oscillator", Phys. Rev. Lett. 478, 1349 (1981).

CHAPTER 4.

[1] Manneville, P., and Pomeau, Y., "Intermittency and the Lorenz Model", Phys. Lett. 75A, 1 (1979); "Different Ways to Turbulence in Dissipative Dynamical Systems", Physica 1D, 219 (1980).

[2] Pomeau, Y., and Manneville, P., "Intermittent Transition to Turbulence in Dissipative Dynamical Systems", Commun. Math. Phys. 74, 189 (1980).

[3] Meirovitch, L., Methods of Analytical Dynamics (McGraw-Hill, New York, 1970).

[4] Hirsch, J.E., Hubermann, J.E., and Scalapino, D.J., "Theory of Intermittency", Phys. Rev. 25A, 519 (1981).

[5] Bergé, P., Dubois, M., Manneville, P., and Pomeau, Y., "Intermittency in Rayleigh-Bénard Convection", J. Phys. (Paris) Lett. 41, L-341 (1980).

[6] Dubois, M., Rubio, M.A., and Bergé, P., "Experimental Evidence of Intermittencies Associated with a Subharmonic Bifurcation", Phys. Rev. Lett. 51, 1446 (1983).

PERTURBATIVE METHODS FOR HAMILTONIAN MAPS

G. Turchetti

Dipartimento di Fisica, Università di Bologna
INFN, Sezione di Bologna
Bologna, Italy

The Hamiltonian dynamics of integrable systems is reviewed in order to explain the perturbative approach to quasi-integrable systems, both in the nonresonant and resonant cases. After defining Hamiltonian maps in the plane, the main results on the qualitative behavior of the orbits are summarized. The normal forms near an elliptic fixed point for polynomial maps are defined and algorithms for the related Birkhoff series are given. In the resonant case the construction of the interpolating Hamiltonians is considered and the geometric properties of the orbits are discussed. The numerical results obtained from the perturbation series for the nonresonant quadratic map allow a heuristic analysis of the asymptotic properties based on the contributions of the leading resonances. A different mechanism leading to convergence for the series which conjugate with circles single invariant curves with diophantine frequency is discussed. The Siegel problem, which exhibits the same convergence mechanism, is discussed and explicit KAM estimates, leading to accurate results, are presented.

INTRODUCTION

In physics, it is customary to use perturbative methods to solve a problem which is close to another whose exact solution is known. Early examples are provided by celestial mechanics, where the orbits of planets are computed treating their mutual interactions as small perturbations to the periodic motion in the central field of the sun. When the perturbation is small, the first terms of the series defining the solution will exhibit a fast decrease and provide accurate results. However, the study of the *convergence* properties is crucial because it can provide information on the analytic structure of the solution and its qualitative behavior. If the series is asymptotic, the a priori estimates on the optimal truncation order are relevant also for high-order numerical computations.

For the continuous and discrete dynamical systems the analysis of perturbation series has been relevant in understanding the geometry of the orbits including the *transition to chaos.* It is interesting in this respect to recall what Poincaré [14] said almost one century ago

Il y a entre les géomètres et les astronomes une sorte de malentendu au sujet de la signification du mot convergence.

Les géomètres, préoccupés de la parfaite rigueur et souvent trop indifférent à la longueur des calculs inextricables dont il conçoivent la possibilité, sans songer à les entreprendre effectivement, disent qu'une série est convergente quand la somme des termes tend vers une limite déterminée, quand même les premiers termes diminueraient très lentement. Les astronomes, au contraire, ont coûtume de dire qu'une série converge quand les vingt premiers termes, par example, diminuent très rapidement, quand même les termes suivants devraient croître indéfiniment.

Ainsi pour prendre un example simple considérons les deux séries qui ont pour terme général

$$\frac{1000^n}{1.2\ldots n} \qquad \frac{1.2\ldots n}{1000^n}$$

Les géomètres diront que la première série converge mais ils regarderont la seconde comme divergente.....

Les astronomes, au contraire, regarderont la première série comme divergente et la seconde comme convergente......

Les deux règles sont légitimes: la première dans les recherches théoriques; la seconde dans les applications numériques. Toutes les deux doivent régner, mais dans deux domaines séparés dont il importe de bien connaître les frontières.

The convergence analysis performed on the canonical perturbation series for Hamiltonian systems led Poincaré to the discovery that the *integrable systems*, whose geometric structure had been decribed by Liouville, was indeed exceptional since any system obtained by adding a generic perturbation had no uniform first integral of motion beyond the Hamiltonian itself. He was the first to suggest the investigation of maps associated to Hamiltonian flows and in a celebrated theorem first gave a description of the rise of deterministic chaos. The geometric structure of the orbits of systems close to integrable was fully understood only much later with the works of Siegel, [40] Kolmogorov, [7] Arnold, [8] Moser [25] and Nekhoroshev. [9]

Hamiltonian maps, called also symplectic maps, have been studied as independent models and today many nonlinear physical systems are modelled directly by maps. The maps have an obvious advantage for direct numerical computation of orbits and they can be theoretically analyzed just as continuous systems.

The aim of these lectures is to describe the behavior of an area preserving map, considering its fixed points and conjugating it with a normal form in a neighborhood. The geometry of orbits becomes transparent: for the elliptic fixed points the orbits of nonresonant normal forms are circles, while the orbits of resonant normal forms are usually similar to those of a pendulum. The conjugation is achieved by series expansions whose construction is carefully described; since these series have a global character, being defined on open neighborhoods of the fixed point, they cannot converge, and both the divergence mechanism and the asymptotic properties are analyzed.

A local conjugation of a nonresonant invariant curve with a circle is also considered, showing with heuristic arguments that the associated series are convergent, the divergence point corresponding to the break-up of the curve.

The plan of these lectures is the following:

In Section 1, we review Hamiltonian systems: for integrable Hamiltonians the resonance pattern in action space is described and the perturbative solution of the conjugation problem is considered for anharmonic oscillators following the recent exhaustive analysis carried out by Benettin, Galgani, Giorgilli and Gallavotti. [10,11]

In Section, 2 the area preserving maps are introduced: the integrable case is described and the instability, under perturbation, of the resonant circles and the rise of chaos is scketched.

In Section 3, we introduce the Birkhoff series for polynomial maps (with a stable fixed point at the origin which are the analog of the anharmonic oscillators). The construction is achieved both in the nonresonant and the resonant case, and the interpolating Hamiltonians are introduced. The extension to symplectic maps of arbitrary dimensionality is considered.

In Section 4, we consider the numerical results obtained for the Birkhoff series and report an argument, based on the dominant resonance contributions, which explains the asymptotic properties of the series. The numerical results for the conjugation of a single nonresonant invariant curve are also reported and the mechanism leading to convergence is described.

In Section 5, another conjugation problem of nonresonant invariant curves with circles, the Siegel problem, is described. The map is in this case analytic rather than area preserving, but the essential features are retained by the conjugation equations allowing an easier comparison between the perturbative results and the estimates obtained from a partial optimization of the KAM method.

1. HAMILTONIAN SYSTEMS

1.1. Hamiltonian flows

A Hamiltonian system is defined by a *phase space* we assume to be $\mathbf{R}^n$ and a function $H : \mathbf{R}^{2n} \to \mathbf{R}$. Denoting by

$$\mathbf{x} = (q_1, \dots, q_n, p_1, \dots, p_n)$$

a point in phase space, the solutions of Hamilton's equations,

$$\dot{q}_i = \frac{\partial H}{\partial p_i} \qquad \dot{p}_i = -\frac{\partial H}{\partial q_i}, \qquad i = 1, \dots, n, \tag{1.1}$$

define a one parameter family of transformations of $\mathbf{R}^{2n}$ into itself called *Hamiltonian flow.*

Introducing the matrix J defined by

$$J = \begin{pmatrix} 0 & I \\ -I & 0 \end{pmatrix},$$

where I is the identity $n \times n$ matrix, Hamilton's equations can be written as

$$\dot{\mathbf{x}} = \mathbf{v}(\mathbf{x}), \qquad \mathbf{v} = J\frac{\partial H}{\partial \mathbf{x}}.$$

Since the vector field $\mathbf{v}$ is solenoidal, $\operatorname{div} \mathbf{v} = 0$, the Hamiltonian flow preserves volumes in phase space (*Liouville's theorem*).

The functions defined in phase space are called *dynamical variables*: the *Poisson's braket* of two dynamical variables A and B is defined by

$$[A, B]_{\mathbf{x}} = \frac{\partial A}{\partial \mathbf{x}} J \frac{\partial B}{\partial \mathbf{x}} \equiv \sum_{l=1}^{n} \left[\frac{\partial A}{\partial q_l}\frac{\partial B}{\partial p_l} - \frac{\partial A}{\partial p_l}\frac{\partial B}{\partial q_l} \right]. \tag{1.2}$$

The trasformations of coordinates in phase space which preserve Hamiltonian flows are called *canonical* or *symplectic*. A trasformation is canonical if and only if

i) the Poisson brakets are invariant,

$$[A, B]_{\mathbf{x}} = [A, B]_{\mathbf{X}} \qquad \forall A, B\,,$$

ii) any one of the following differential forms is exact:

$$\mathbf{p}d\mathbf{q} - \mathbf{P}d\mathbf{Q}, \qquad \mathbf{p}d\mathbf{q} + \mathbf{Q}d\mathbf{P}, \qquad -\mathbf{q}d\mathbf{p} - \mathbf{P}d\mathbf{Q}, \qquad -\mathbf{q}d\mathbf{p} + \mathbf{Q}d\mathbf{P}.$$

Letting $F(\mathbf{q},\mathbf{P})$ be the *generating* function whose differential is the second form written above, the canonical transformation is implicitly defined by

$$\mathbf{P} = \frac{\partial F}{\partial \mathbf{q}}, \qquad \mathbf{Q} = \frac{\partial F}{\partial \mathbf{P}}. \tag{1.3}$$

Since $F = \mathbf{q}\cdot\mathbf{P}$ is the identity transformation, any one parameter family of canonical transformations close to the identity can be written

$$F = \mathbf{q}\cdot\mathbf{P} + \sum_{l\geq 1} \epsilon^l F_l(\mathbf{q},\mathbf{P}). \tag{1.4}$$

The Hamiltonian flows are one parameter families of canonical transformations.

A linear transformation $\mathbf{X} = L\mathbf{x}$ is canonical if and only if L is a symplectic matrix, that is, $\tilde{L}JL = J$. The jacobian M of a canonical diffeomorphism is a symplectic matrix; indeed, letting $M_{ik} = \frac{\partial X_k}{\partial x_i}$ one has

$$\frac{d\mathbf{X}}{dt} = \tilde{M}\frac{d\mathbf{x}}{dt} = \tilde{M}JM\frac{\partial H}{\partial \mathbf{X}} = J\frac{\partial H}{\partial \mathbf{X}}.$$

The one-parameter families of canonical diffeomorphisms can be explicitly written using Lie series. Defining the Lie derivative D_η with repect to a dynamical variable $\eta(\mathbf{x})$ according to

$$(D_\eta f)(\mathbf{x}) = [f,\eta]_{\mathbf{x}}$$

and the exponential of D_η as the Lie series

$$e^{\epsilon D_\eta} f = \sum_{l=0}^{\infty} \frac{\epsilon^l}{l!} D_\eta^l f,$$

the transformation $\mathbf{X} = e^{\epsilon D_\eta}\mathbf{x}$ is canonical and corresponds to a Hamiltonian flow with Hamilton's function $\eta(\mathbf{x})$ since one proves that

$$\frac{d\mathbf{X}}{d\epsilon} = J\frac{\partial \eta}{\partial \mathbf{X}}.$$

We have considered only time independent Hamiltonians because any time dependence can be absorbed by introducing an additional degree of freedom, that is, $H(\mathbf{q},\mathbf{p},t)$ has the same equations of motion as $H' = p_{n+1} + H(\mathbf{q},\mathbf{p},q_{n+1})$.

1.2. Integrable systems

A system is integrable if it has n independent integrals of motion $\psi_k(\mathbf{x})$, $k = 1,\ldots,n$, in involution and the energy surface $H(\mathbf{x}) = E$ is compact. In this case the geometry of the orbits is simple because the energy surface is foliated into *invariant tori*, that is, for any initial condition the motion is equivalent to uniform rotations on n distinct circles. Indeed, a theorem by Liouville and Arnold [2] states that:

Theorem. If $\psi_1(\mathbf{x}), \ldots, \psi_n(\mathbf{x})$ are

i) independent: $\text{grad}\psi_1, \ldots, \text{grad}\psi_n$ linearly independent;

ii) in involution: $[\psi_i, \psi_k] = 0 \qquad i, k = 1, \ldots, n$;

iii) the manifold $M_\gamma \equiv \{\mathbf{x} : \psi_i(\mathbf{x}) = \gamma_i\}$ is compact and connected;

iv) the Hamiltonian is $H(\psi_1, \ldots, \psi_n)$ so that the ψ_k are constants of motion;

then M_γ is diffeomorphic to the torus $\mathbf{T}^n$, and letting φ denote the angular coordinates on $\mathbf{T}^n$ the conjugate variables j are defined by the mapping $\mathrm{j} = \mathrm{j}(\gamma)$ obtained by quadratures.

It is therefore equivalent to define as *integrable* a system whose phase space is $\mathbf{R}^n \times \mathbf{T}^n$ and whose Hamiltonian is $\hat{H}(\mathrm{j})$. In order to illustrate the theorem we consider a Hamiltonian

$$H = K(\psi_1(q_1, p_1), \ldots, \psi_n(q_n, p_n)) \tag{1.5}$$

for which the Hamilton-Jacobi equation is separable. We further assume that the motion in each plane q_i, p_i is periodic, which is equivalent to demanding that each curve $\psi_i(q_i, p_i) = \gamma_i$ and consequently M_γ is compact. If q_i is a cartesian coordinate then $\psi_i(q_i, p_i) = \gamma_i$ is a closed curve, if q_i is an angle then $\psi_i(q_i, p_i)$ is 2π periodic in q_i. Two typical examples are given by harmonic oscillators and rotating pendulums (see Fig. 1):

$$H = \frac{1}{2}\sum_{l=1}^{n}(p_l^2 + \omega_l^2 q_l^2), \qquad H = \frac{1}{2}\sum(p_l^2 - g\cos\varphi_l).$$

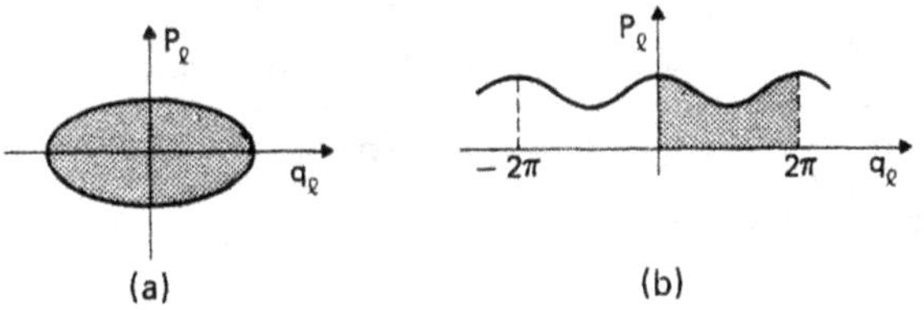

Figure 1

The *Hamilton-Jacobi* equation defines the canonical transformation $(\mathbf{q}, \mathbf{p}) = \mathcal{C}(\mathbf{Q}, \mathbf{P})$ which brings the Hamiltonian to a form independent of the coordinates

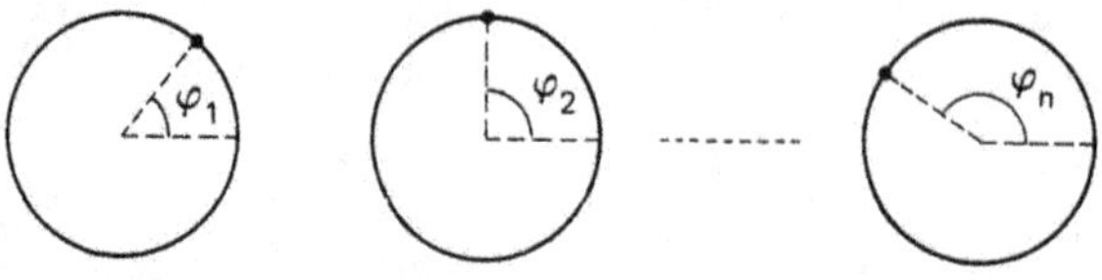

Figure 2

$H(C(\mathbf{Q},\mathbf{P})) = \hat{H}(\mathbf{P})$. Using a generating function $F(\mathbf{q},\mathbf{P})$ to achieve the transformation C, the Hamilton-Jacobi equation reads

$$H\left(\mathbf{q}, \frac{\partial F}{\partial \mathbf{q}}(\mathbf{q},\mathbf{P})\right) = \hat{H}(\mathbf{P}). \tag{1.6}$$

The new Hamiltonian is to a large extent arbitrary and so is the transformation C. Indeed, separating the variables $F(\mathbf{q},\mathbf{P}) = F_1(q_1,P_1) + \cdots + F_n(q_n,P_n)$ the Hamilton-Jacobi equation is solved by

$$\psi_l(q_l, \frac{dF_l}{dq_l}) = \gamma_l(\mathbf{P}), \quad l = 1,..,n, \qquad \hat{H}(\mathbf{P}) = K(\gamma(\mathbf{P})),$$

where the functions γ are arbitrary. Since the areas delimited by the orbits in each phase plane q_l, p_l are invariant, one introduces the *action variables*

$$j_l = \frac{1}{2\pi}\oint p_l dq_l$$

as new momenta and proves that the canonically conjugated coordinates φ are angles. The transformation $(\mathbf{q},\mathbf{p}) \to (\varphi, \mathrm{j})$ is obtained from a generating function

$$f(\mathbf{q},\mathrm{j}) = F(\mathbf{q},\mathbf{P}(\mathrm{j})),$$

where $\mathbf{P} = \mathbf{P}(\mathrm{j})$ is the inverse of the map $\mathrm{j} = \mathrm{j}(\mathbf{P})$ given by

$$j_l(\mathbf{P}) = \frac{1}{2\pi}\oint \frac{\partial F_l}{\partial q_l}(q_l,\mathbf{P})dq_l.$$

The variation of φ_l when a cycle is described in the (q_k,p_k) plane is $2\pi\delta_{kl}$ and the Hamiltonian becomes

$$h(\mathrm{j}) = H(\mathbf{P}(\mathrm{j})).$$

We have therefore succeded in finding a coordinate system in phase space such that the motion becomes

$$\begin{cases} \varphi(t) = \omega(\mathrm{j})t + \varphi(0), \\ \mathrm{j}(t) = \mathrm{j}(0), \end{cases} \qquad \omega(\mathrm{j}) = \frac{\partial h}{\partial \mathrm{j}}, \tag{1.7}$$

that is, uniform rotations with frequencies $\omega(\mathrm{j})$ (see Fig. 2). The manifold M_γ is therefore the product of n circles, that is a torus $\mathbf{T}^n$ and the phase space is $\mathbf{R}^n \times \mathbf{T}^n$. A constant energy surface is foliated into invariant tori.

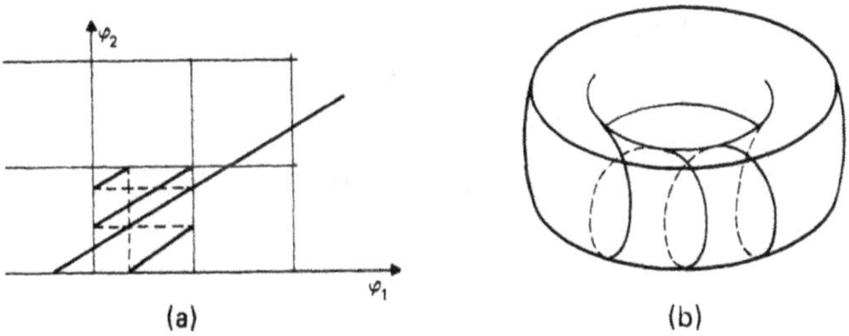

Figure 3

We recall that $\mathbf{T}^n = \mathbf{R}^n/\mathbf{Z}^n$ and consequently the trajectories on a torus can be represented as trajectories on $\mathbf{R}^n$, provided that all the points translated by integer multiples of 2π are identified (see Fig. 3).

For each degree of freedom (j_l, φ_l) we have a manifold $\mathbf{R} \times \mathbf{T}$ which can be represented as a cylinder or a family of concentric circles (treating j_l and φ_l as polar canonical coordinates in a plane $\mathbf{R}^2$ whose cartesian coordinates are $x_l = \sqrt{2j_l}\cos\varphi_l$, $y_l = \sqrt{2j_l}\,\sin\varphi_l$) or as a plane where the points horizontally translated by multiple integers of 2π are identified ($\mathbf{R} \times \mathbf{T} = \mathbf{R}^2/\mathbf{Z}$) (see Fig. 4).

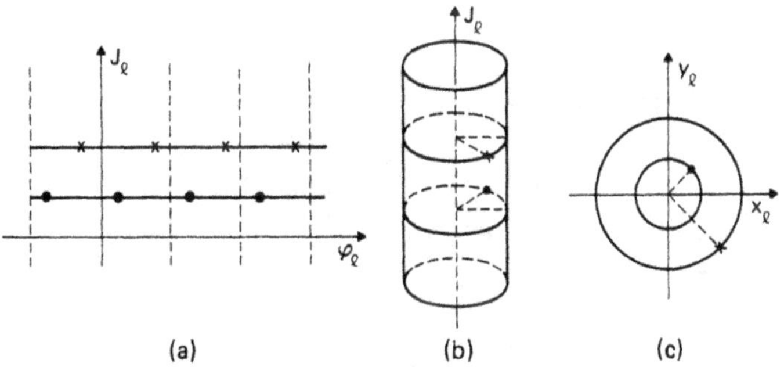

Figure 4

1.3. Resonances

The integrable Hamiltonians are divided into two classes: *isochronous* and *anisochronous*. The former have a constant frequency

$$h = \omega \mathbf{j}$$

and correspond to harmonic oscillators (the transformation is $q_l = \sqrt{\frac{2j_l}{\omega_l}}\sin\varphi_l$, $p_l = \sqrt{2j_l\omega_l}\cos\varphi_l$); the latter have a variable frequency depending on the action, and a typical Hamiltonian is

$$h = \sum_{l=1}^{n} \frac{j_l^2}{2I_l},$$

which corresponds to a system of free rotators. If the *nondegeneracy* condition

$$\det A \neq 0, \qquad A_{kl} = \frac{\partial^2 h}{\partial j_k \partial j_l} = \frac{\partial \omega_k}{\partial j_l},$$

is satisfied, then a bijection between frequencies and actions exists.

A frequency can be *nonresonant* or *resonant*. A frequency ω is said to be *nonresonant* if

$$\omega \cdot \mathbf{k} \neq 0 \quad \text{for} \quad \forall \mathbf{k} \in \mathbf{Z}^n \backslash 0. \tag{1.8}$$

It is said to satisfy a resonance condition of order r if there is subspace $W_r \subset \mathbf{Z}^n$ of dimension r such that

$$\begin{cases} \omega \cdot \mathbf{k} = 0 & \forall \mathbf{k} \in W_r \subset \mathbf{Z}^n, \\ \omega \cdot \mathbf{k} \neq 0 & \forall \mathbf{k} \notin W_r \subset \mathbf{Z}^n. \end{cases} \tag{1.9}$$

A torus $\mathbf{j}$ is said to be resonant or nonresonant if $\omega(\mathbf{j})$ is resonant or nonresonant.

This distinction is of basic importance. Indeed, consider a dynamical variable f defined on $\mathbf{T}^n$ that is a periodic function on $\mathbf{R}^n$: $f(\varphi) = f(\varphi + 2\pi\mathbf{k}), \mathbf{k} \in \mathbf{Z}^n$ and define the space average according to

$$\langle f \rangle_S = (2\pi)^{-n} \int_{\mathbf{T}^n} f(\varphi) d\varphi_1, ..., d\varphi_n$$

and the time average on an orbit by

$$\langle f \rangle_T = \lim_{T \to +\infty} \frac{1}{T} \int_0^T f(\varphi(t)) dt,$$

where $\varphi(t) = \omega t + \varphi(0)$.
We remark that if

$$f(\varphi) = \sum_{\mathbf{k} \in Z^n} f_{\mathbf{k}} e^{i\mathbf{k} \cdot \varphi}$$

is the Fourier expansion of $f(\varphi)$, then

$$\langle f \rangle_S \equiv f_0.$$

Theorem. If ω is nonresonant the space and time averages are equal. The motion is said to be ergodic. The trajectory is dense on the torus $\mathbf{T}^n$. If ω satisfies a resonance condition of order r then the trajectory is no longer dense on $\mathbf{T}^n$, but only on $\mathbf{T}^{n-r}$.

The theorem is trivial to prove if f is a trigonometric polynomial, since in this case $\langle f\rangle_T = a_0 + \lim_{T\to\infty}\frac{1}{T}\left[F(\varphi(T)) - F(\varphi(0))\right]$, where $F(\varphi) = \frac{1}{i}\sum_{\mathbf{k}\neq 0}\frac{f_{\mathbf{k}}}{\omega\cdot\mathbf{k}}e^{i\mathbf{k}\cdot\varphi}$ is bounded and satisfies the linear partial differential equation $\omega\cdot\frac{\partial F}{\partial\varphi} = f - \langle f\rangle_S$

Since resonant and nonresonant frequencies imply a totally different behavior of the orbits we briefly examine their geometric structure according to [11]. For $W_r \subset \mathbf{Z}^n$ of dimension r, we define a resonant surface $\mathcal{M}(W_r)$ in the frequency space by

$$\mathcal{M}(W_r) = \{\omega | \omega\in\mathbf{R}^n;\ \omega\cdot\mathbf{k} = 0 \quad \mathbf{k}\in W_r;\ \omega\cdot\mathbf{k}\neq 0 \quad \mathbf{k}\notin W_r\}.$$

The set $\mathcal{M}(W_r)$ is the intersection of all the planes $\omega\cdot\mathbf{k}_i = 0$ for $i = 1,\ldots,r$, where $\mathbf{k}_i$ are a basis in W_r, deprived of the intersection with any other resonant plane $\omega\cdot\mathbf{k} = 0$ where $\mathbf{k}\notin W_r$. A resonant structure or *block* of order r $\Re_r$ will be the union of all the resonant surfaces

$$\Re_r = \bigcup_{W_r\subset Z^n} \mathcal{M}(W_r).$$

For $n = 2$, we see that each $\mathcal{M}(W_1)$ is a straight line through the origin with rational angular coefficient, deprived of the origin itself (see Fig. 5a). As a consequence, $\Re_1$ is not connected. For $n = 3$, each $\mathcal{M}(W_1)$ is a plane deprived of the intersections with other such planes, since these intersections belong to $\Re_2$ (see Fig. 5b).

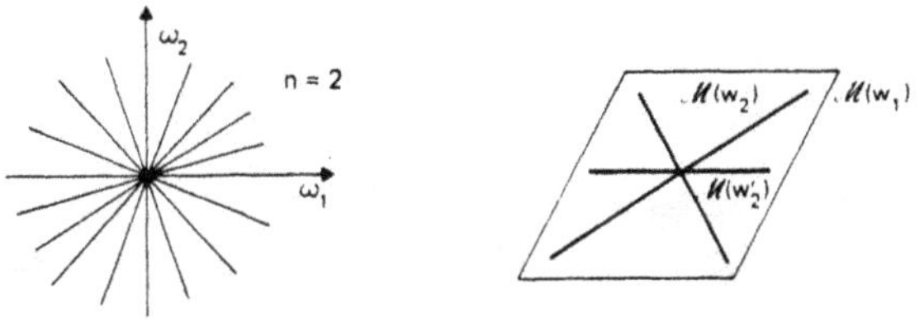

(a) (b)

Figure 5

Letting $\Re_0$ denote the nonresonant surface

$$\Re_0 = \{\omega|\omega\in\mathbf{R}^n; \omega\cdot\mathbf{k}\neq 0 \quad \forall k\in\mathbf{Z}^n\backslash 0\}$$

and $\Re_n$ the set containing the single point $\omega = 0$, the space of frequencies $\mathbf{R}^n$ is partitioned into resonant blocks

$$\mathbf{R}^n = \bigcup_{r=0}^{n}\Re_n.$$

The space of resonant frequencies $\bigcup_{r=1}^{n} \Re_r$ is dense in $\mathbf{R}^n$, but its intersection with any compact set A has zero Lebesgue measure, while the measure of $\Re_0 \cap A$ is equal to the measure of A.

Finally, for any asynchronous nondegenerate system the same picture holds, up to a continuous deformation in the action space. The geometry of resonant sets in a constant energy surface $\mathcal{E}_E$ is also easily described. Considering, for instance, $h = \sum_{l=1}^{n} \frac{j_l^2}{2I_l}$, we see that $\mathcal{E}_E$ is an ellipsoid. For $n = 2$, the intersection $\Re_1 \cap \mathcal{E}_E$ is a disconnected dense set of measure 0 (see Fig. 6a). For $n = 3$, the set $(\Re_1 \cup \Re_2) \cap \mathcal{E}_E$ is given by all the ellipses obtained by intersecting the ellipsoid with the planes $\omega \cdot \mathbf{k} = 0$, $\Re_2 \cap \mathcal{E}_E$ being given by the intersection of all these ellipses. The sets $\Re_1 \cap \mathcal{E}_E$ and $\Re_2 \cap \mathcal{E}_E$ are disconnected, but their union, which covers the ellipsoid as a spider-web, is obviously connected (see Fig. 6b).

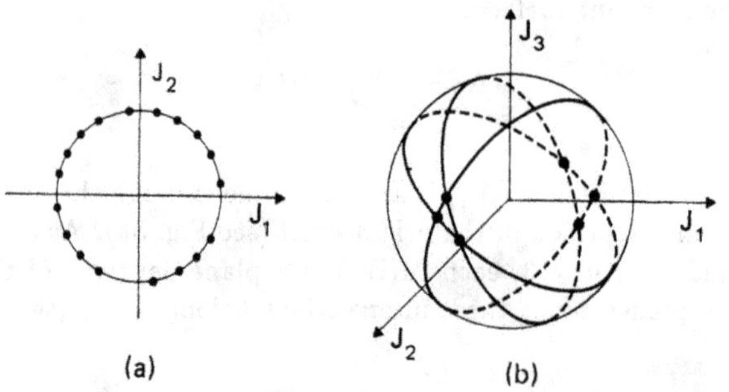

Figure 6

1.4. Quasi-integrable systems

We consider a system obtained by perturbing an integrable one,

$$h(\mathrm{j}, \varphi; \epsilon) = H_0(\mathrm{j}) + \epsilon V(\mathrm{j}, \varphi), \tag{1.10}$$

and ask whether a transformation $(\mathrm{j}, \varphi) = \mathcal{C}(\mathbf{J}, \mathbf{\Phi}; \epsilon)$ can be found such that the system takes an explicitly integrable form,

$$h(\mathcal{C}(\mathbf{J}, \mathbf{\Phi}; \epsilon); \epsilon) = \hat{H}(\mathbf{J}; \epsilon). \tag{1.11}$$

Already Poincaré [14] had shown that if h is analytic and V has sufficiently many nonvanishing Fourier components, no analytic solution $\mathcal{C}$ exists. Fifty years later Kolmogorov, Arnold, and Moser (KAM) proved that only the strongly nonresonant tori, i.e., with a frequency $\omega(\mathrm{j}) = \frac{\partial H_0(\mathrm{j})}{\partial \mathrm{j}}$ satisfying a diophantine condition

$$|\omega \cdot \mathbf{k}|^{-1} \leq \gamma ||\mathbf{k}||^{\mu}, \tag{1.12}$$

are preserved by sufficiently small perturbations. The resonant tori are generically unstable and are destroyed by any perturbation by a mechanism described by Poincaré and Birkhoff in a celebrated theorem that will be described later.

This result is not surprising, due to the different nature of the orbits whose closure is the torus $\mathbf{T}^n$ only in the nonresonant case. The numerical analysis performed by integrating the Hamilton's equations and constructing the Poincaré section confirm that when the perturbation is increased, more and more irregular orbits appear.

Since uniform solutions in phase space cannot be found for the Hamilton-Jacobi Eq. (1.6), a more modest program consists in looking for *perturbative* solutions. We consider a canonical tranformation C_N, which is a polynomial in ϵ of order N satisfying

$$h(C_N(\mathbf{J},\mathbf{\Phi};\epsilon);\epsilon) = \sum_{s=0}^{N} H_s(\mathbf{J})\epsilon^s + \epsilon^{N+1}R_N(\mathbf{J},\mathbf{\Phi};\epsilon). \tag{1.13}$$

A necessary and sufficient condition for this equation to have a solution is that $\omega(\mathbf{J}) = \frac{\partial H_0(\mathbf{J})}{\partial \mathbf{J}}$ is nonresonant. If ω is resonant, $\mathbf{k}\cdot\omega = 0$, $\mathbf{k} \in W_r \subset \mathbf{Z}^n$, then $H_s(\mathbf{J})$ must be replaced by

$$H_s(\mathbf{J},\mathbf{\Phi}) = \sum_{\mathbf{k}\in W_r} H_{s,\mathbf{k}}(\mathbf{J})e^{i\mathbf{k}\cdot\mathbf{\Phi}}. \tag{1.14}$$

The consequence is that only $n-r$ approximate first integrals of motion exist. Indeed, letting $\mathbf{m}$ be any one of the $n-r$ linearly independent vectors in the subspace of $\mathbf{Z}^n$ orthogonal to W_r, one has

$$\frac{d}{dt}(\mathbf{m}\cdot\mathbf{J}) = -i\sum_{\mathbf{k}\in W_r} \mathbf{m}\cdot\mathbf{k}H_{s,\mathbf{k}}e^{i\mathbf{k}\cdot\mathbf{\Phi}} + O(\epsilon^{N+1}) = O(\epsilon^{N+1}).$$

For isochronous systems, the frequency is fixed and determines the perturbation scheme to be used. For anisochronous systems, the choice depends pointwise on phase space. However, when V is a trigonometric polynomial or if the ultraviolet cutoff is used, then at any order only a finite (and increasing) number of Fourier components is present and a smoother partitioning of frequency and action spaces into resonant manifolds can be used. The presence of a finite number of resonances allows the introduction of thick manifolds, according to a scheme introduced by Nekoroshev, to investigate the *time regime* under which bounds to the variation of the actions could be given. For Nth order perturbation theory, the actions are quasi-constant in a time interval $\sim \epsilon^{-N}$. Indeed, from the equations of motion

$$\frac{d\mathbf{J}}{dt} = -\epsilon^{N+1}\frac{\partial R_N}{\partial \mathbf{\Phi}}$$

it follows that if $R_N \in C^1$ in a domain $D = S(\rho)\times\mathbf{T}^n$, where S denotes a sphere of $\mathbf{R}^n$, then if $M = ||\frac{\partial R_N}{\partial \mathbf{\Phi}}||_D$ and if $\mathbf{J}(0) \in S(\rho-\delta)$, where $\delta > \epsilon M$, the following estimate holds:

$$||\mathbf{J}(t) - \mathbf{J}(0)|| \leq \epsilon^{N+1}tM \leq \epsilon, \qquad \text{if} \quad t \leq \epsilon^{-N}M^{-1}.$$

If the sequence of remainders converged to zero, the actions would be constant for arbitrarily long times in contrast with the KAM theorem. Indeed, it has been proved in [10] and [11] that the series are *asymptotic*. If $H_0(\mathrm{j}) = \omega \cdot \mathbf{J}$ with ω diophantine (see 1.12), then

$$||R_N||_D \leq A(BN)^{\beta N},$$

where A, B, β are suitable constants depending on $||V||$, n, γ. This implies that the remainder behaves as $\epsilon^{N+1}||R_N||_D \sim x^N N^{\beta N}$, where $x = \epsilon B^{\beta}$, and has a minimum for $\overline{N} = e^{-1}x^{-1/\beta}$, and correspondingly

$$\epsilon^{\overline{N}}||R_{\overline{N}}||_D \leq A \exp(-\beta \overline{N}) = A \exp\left(-\frac{\epsilon^{-1/\beta}}{eB}\right).$$

It follows therefore that the optimal estimate of the time regime for which $||\mathbf{J}(t) - \mathbf{J}(0)|| < \epsilon$ is (letting A' be the constant corresponding to the derivative of R_N)

$$t \leq A'^{-1} \exp\left(\frac{\epsilon^{-1/\beta}}{eB}\right).$$

To conclude this analysis, we write the recursion relation for the perturbation expansion when $H_0(\mathrm{j}) = \omega \cdot \mathrm{j}$ using the generating function method $F = \varphi \cdot \mathbf{J} + \sum_{s=1}^{n} \epsilon^s F_s(\varphi, \mathbf{J})$ to define C_n. Then at order s we have

$$\omega \cdot \frac{\partial F_s(\varphi, \mathbf{J})}{\partial \varphi} + Q_s(\varphi, \mathbf{J}) = H_s(\mathbf{J}), \tag{1.15}$$

where Q_s depend on $F_1, \ldots, F_{s-1}$. If $Q_{s,\mathbf{k}}$ are the Fourier coefficients of Q_s, we have

$$H_s(\mathbf{J}) = Q_{s,0}(\mathbf{J}), \qquad F_s = i \sum_{\mathbf{k} \neq 0} \frac{Q_{s,\mathbf{k}}}{\omega \cdot \mathbf{k}} e^{i\mathbf{k} \cdot \varphi},$$

provided that ω is not resonant. If ω is resonant with respect to $W_r \subset \mathbf{Z}^n$, then in equation (1.15) $H_s(\mathbf{J})$ is replaced by $H_s(\mathbf{J}, \varphi)$, given by (1.14) and one obtains

$$H_s(\mathbf{J}, \varphi) = \sum_{\mathbf{k} \in W_r} Q_{s,\mathbf{k}}(\mathbf{J}) e^{i\mathbf{k} \cdot \varphi}, \qquad F_s = i \sum_{\mathbf{k} \notin W_r} \frac{Q_{s,\mathbf{k}}}{\omega \cdot \mathbf{k}} e^{i\mathbf{k} \cdot \varphi}.$$

If V is not a trigonometric polynomial, the convergence of the series is insured if V is analytic in φ so that its Fourier coefficients have an exponential decrease and ω satisfies a diophantine condition (1.12). We recall also that the measure of the set of the frequencies satisfying this condition is nonzero only if we choose $\mu > n$.

1.5. The origin of islands and chaotic motions

The behavior of a strongly nonresonant torus when a perturbation is switched on is simple: for sufficiently small perturbations the topological structure is preserved, the deformations being described by diffeomorphisms. The fate of resonant

tori is different: they generically blow up under any perturbation and the nearby orbits exhibit a very reach phenomenology. The systems with two degrees of freedom exhibit, in a suitable plane of section, *elliptic points* surrounded by islands and *hyperbolic points* whose stable and unstable manifolds undergo homoclinic intersections leading to chaotic motions. A good description is obtained using resonant perturbation theory in the neighborhood of a resonance.
We consider the Hamiltonian

$$H(\mathrm{j},\varphi) = \frac{j_1^2 + j_2^2}{2} + \epsilon V(\mathrm{j},\varphi)$$

and a resonant action $\mathrm{j}^{\mathrm{res}}$. In this model, frequencies and actions coincide, $\omega = \mathrm{j}$, and the resonance condition is

$$\mathbf{k}_0 \cdot \mathrm{j}^{\mathrm{res}} = 0, \qquad \mathbf{k}_0 = (n, -m),$$

where m and n are relatively prime and all the resonant vectors $\mathbf{k} \in W_1$ are given by $\mathbf{k} = p\mathbf{k}_0, \quad \forall p \in \mathbf{Z}$. To first order, the resonant perturbation theory gives

$$H(\mathrm{j},\varphi) = \hat{H}(\mathbf{J},\mathbf{\Phi}) = \frac{J_1^2 + J_2^2}{2} + \epsilon \sum_{p=0}^{\infty} V_p(\mathbf{J}) \cos(p\mathbf{k}_0 \cdot \mathbf{\Phi}) + \epsilon^2 R(\mathrm{j}, \mathbf{\Phi}; \epsilon),$$

where $V_p = 2V_{p\mathbf{k}_0}$ if $p \neq 0$, $V_0 = V_0$ and R denotes the remainder. The new resonant action is $\mathbf{J}^{\mathrm{res}}$, where $\mathbf{J}^{\mathrm{res}} \cdot \mathbf{k}_0 = 0$ and differs from $\mathrm{j}^{\mathrm{res}}$ by $O(\epsilon)$.

The linear canonical transformation

$$\begin{cases} \Psi_1 = \Phi_1 - \frac{m}{n}\Phi_2 & I_1 = (J_1 - \frac{m}{n}J_2)/(1 + \frac{m^2}{n^2}) \\ \Psi_2 = \frac{m}{n}\Phi_1 + \Phi_2 & I_2 = (\frac{m}{n}J_1 + J_2)/(1 + \frac{m^2}{n^2}) \end{cases}$$

leads to an Hamiltonian depending on only one angle variable:

$$H = (1 + \frac{m^2}{n^2})\frac{1}{2}(I_1^2 + I_2^2) + \epsilon \sum_p \hat{V}(\mathbf{I}) \cos(pn\Psi_1) + \epsilon^2 R$$

The resonant action becomes $\mathbf{I}^{\mathrm{res}} = \mathbf{I}(\mathbf{J}^{\mathrm{res}})$ and it follows that $I_1^{\mathrm{res}} = 0$. In order to analize the behavior in a neighborhood of the resonant actions, we consider the following transformation which leaves the equations of motion invariant:

$$\begin{cases} I_1 = \sqrt{\epsilon} I_1', & \Psi_1 = \Psi_1', & H = \epsilon H', \\ I_2 = I_2^{\mathrm{res}} + \sqrt{\epsilon} I_2', & \Psi_2 = \Psi_2', & t = \frac{t'}{\sqrt{\epsilon}}. \end{cases}$$

The new Hamiltonian is given by

$$H' = (1+\frac{m^2}{n^2})\frac{I_1'^2}{2} + \sum_p \hat{V}(\mathbf{I}^{\mathrm{res}}) \cos(np\Psi_1) + (1+\frac{m^2}{n^2})(\frac{I_2^{\mathrm{res}}}{\sqrt{\epsilon}} I_2' + I_2'^2) + \sqrt{\epsilon} R'. \quad (1.16)$$

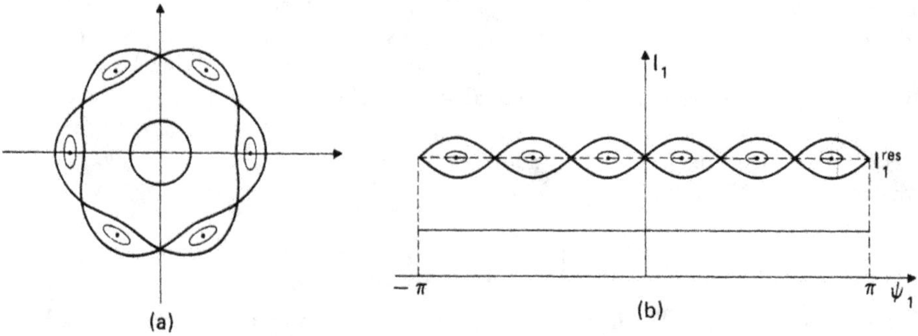

Figure 7

It is evident that there is a fast variable Ψ_2' which evolves with a frequency $\sim \frac{1}{\sqrt{\epsilon}}$ while the corresponding action is frozen, $I_2' \sim \sqrt{\epsilon t'}$ .The other variable Ψ_1' is slow, evolving with a frequency of order 1. If the coefficients V_p decrease rapidly with p so that we can neglect all of them but the first one, then the motion is the same as for a pendulum (see Fig. 7).

As a consequence, the orbits around the elliptic points of the pendulum and the separatrices describe the *islands* observed in numerical experiments. However, at this stage no chaotic motion is allowed. The distance of the separatrices of the pendulum $\Delta = 4[V_1(\mathbf{I}^{\rm res})/(1+\frac{m^2}{n^2})]^{1/2}$ gives an estimate of the width of the islands.

If one retains the next terms up to order ϵ, then a coupling with the fast variable Ψ_2' is present, and since $\Psi_2' \sim I_2^{\rm res}(1+\frac{m^2}{n^2})\frac{t'}{\sqrt{\epsilon}}$ the *slow variables* (I_1', Ψ_1') evolve as a forced pendulum which exhibits chaotic motion according to Melnikhov's theorem on homoclinic intersections. [12] We observe that the resonant normal form can be calculated to higher orders with a correspondingly decreasing remainder. Indeed, as proved by Nekhoroshev, the remainder can be made exponentially small $\sim e^{-B\epsilon^{-\beta}}$ but it still remains to be proved that the chaotic motions persist.

2. AREA PRESERVING MAPS

2.1. Poincaré section

Given any system with two degrees of freedom, an area preserving map is obtained associating to any orbit in a constant energy surface a sequence of points in which the orbit intersects a given plane with the *same orientation* (if $\mathbf{n}$ is the normal to the plane and $\dot{\mathbf{x}} = \mathbf{v}$ are the equations of motion $\mathbf{n} \cdot \mathbf{v}$ must have a constant sign; see Fig. 8).

For a periodic time dependent system with one degree of freedom, the map

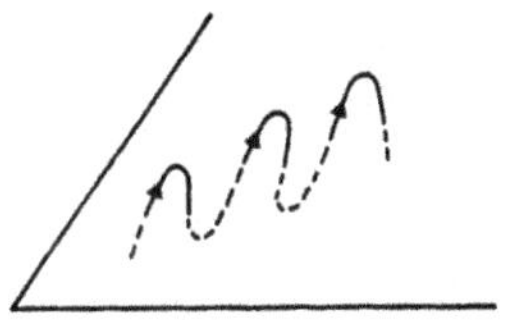

Figure 8

is obtained by taking the stroboscopic images of the system. Letting

$$H(q,p,t) = H(q,p,t+T)$$

and letting $S_t(q_0,p_0)$ denote the Hamiltonian flow, then the sequence $(q_n,p_n) = S_{nT}(q_0,p_0)$ for $n = 0,1,\ldots$ defines an area preserving map. Indeed, one has

$$\frac{\partial(q_n,p_n)}{\partial(q_0,p_0)} = [q_n,p_n]_{q_0,p_0} = 1.$$

The first example we consider corresponds to the map obtained from a two dimensional *integrable* system with Hamiltonian $H(\mathrm{j})$ and phase space $\mathbf{R}^2 \times \mathbf{T}^2$. An orbit is given by $\varphi(t) = \omega(\mathrm{j})t + \varphi(0)$ and we consider $\varphi_2 = 0 \pmod{2\pi}$ as the plane of section. The intersection times are given by

$$\varphi_2(t_n) = 2\pi n \qquad \Rightarrow \qquad t_n = \frac{2\pi n - \varphi_2(0)}{\omega_2},$$

and since $\dot{\varphi}_2 = \omega_2$ has a constant sign, all the intersections have to be taken and the map turns out to be

$$\begin{cases} \varphi_1(t_n) = 2\pi\frac{\omega_1}{\omega_2}n - \frac{\omega_1}{\omega_2}\varphi_2(0) + \varphi_1(0), \\ j_1(t_n) = j_1(0), \end{cases}$$

and the map on $\mathbf{R} \times \mathbf{T}$, that is, of the *cylinder* into itself, can be written

$$\begin{cases} \varphi_1' = \varphi_1 + \Omega(j_1), \\ j_1' = j_1, \end{cases} \tag{2.1}$$

where $\Omega = 2\pi\omega_1/\omega_2$ is a function of j_1. Indeed, the orbits are taken on a constant energy surface $H(j_1,j_2) = E$, where assuming $\omega_2 \neq 0$ we can invert obtaining $j_2 = j_2(j_1)$.

The intersection of the family of tori into which the surface $H = E$ is foliated with the plane $\varphi_2 = 0$ is a family of circles to which the orbits of the map belong. If ω is nonresonant, then ω_1/ω_2 is *irrational*, and since the orbit of H is dense on $\mathbf{T}^2$ the corresponding orbit of the map is dense on the circle. If ω is resonant, that

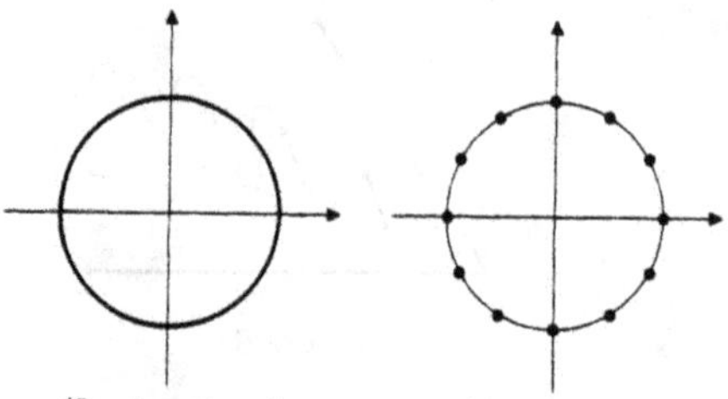

$\pi/2\pi$ irrational (a) $\pi/2\pi$ rational (b)

Figure 9

is $\omega_1/\omega_2 = p/q$, where $p, q \in \mathbf{Z}$ are relatively prime, then the orbits of H, which are closed and not dense on $\mathbf{T}^2$, intersect the section plane q times and the orbits of the map consist of only q points (see Fig. 9).

When $\Omega(j)$ is monotonic (we drop from now on the index 1 of j_1, φ_1), then the inverse $j = j(\Omega)$ exists and we can label the circles with Ω as well as with j; in this case the map is called a *twist map.* We see that j belongs to a compact set $J \subset \mathbf{R}$ because $H = E$ is compact; as a consequence, since the Lebesgue measure of the rationals in any closed interval is zero, if $\Omega(j)$ is a diffeomorphism the measure of the subset of J corresponding to rational frequencies is also zero even though it is *dense.* For a twist map the nonresonant circles alternate with the resonant ones, the latter being dense and of zero Lebesgue measure. According to Moser's theorem,[25] when an integrable twist map is perturbed, only the strongly nonresonant circles are generically preserved by a perturbation analytic and sufficiently small. We call a frequency ω and the corresponding circle $j(\omega)$ strongly nonresonant if it satisfies a diophantine condition

$$|\frac{\Omega}{2\pi} - \frac{p}{q}|^{-1} < \gamma q^{\mu}, \qquad \forall p, q \in \mathbf{Z}, \quad \mu \geq 2. \tag{2.2}$$

The complement of the set of strongly nonresonant circles is dense, but of small measure. For a fixed μ, we consider the set of frequencies which do not satisfy the diophantine condition, that is, $|\Omega/2\pi - p/q| \leq \gamma^{-1}q^{-\mu}$ for $\forall p, q \in \mathbf{Z}$. For fixed q, we consider a set $\mathcal{R}_q$ of q intervals centered at $1/q, 2/q \ldots, q/q$ of length $\gamma^{-1}q^{-\mu}$ so that $\text{meas}(\mathcal{R}_q) \leq \gamma^{-1}q^{-\mu-1}$. By taking the union of $\mathcal{R}_q$, we obtain a set $\mathcal{R}$ whose measure is bounded by

$$\text{meas}(\mathcal{R}) \leq \sum_{q=1}^{\infty} \gamma^{-1}q^{-\mu+1} = \gamma^{-1}\varsigma(\mu - 1),$$

where ς denotes the Riemann zeta function. The measure of the complement $\mathcal{N}$, that is, of the set of the strongly nonresonant frequencies, is bounded by $\text{meas}(\mathcal{N}) = 1 - \text{meas}(\mathcal{R}) \geq 1 - \gamma^{-1}\varsigma(\mu - 1)$, and since γ can be arbitrarily large the measure of $\mathcal{N}$ is arbitrarily close to 1.

Before considering the perturbation of integrable maps we state the expression it takes in cartesian coordinates:

$$x = \sqrt{2j}\cos\varphi, \qquad y = \sqrt{2j}\sin\varphi,$$

that is

$$\begin{pmatrix} x' \\ y' \end{pmatrix} = R\left[\Omega(\frac{x^2+y^2}{2})\right]\begin{pmatrix} x \\ y \end{pmatrix}, \qquad R(\alpha) = \begin{pmatrix} \cos\alpha & -\sin\alpha \\ \sin\alpha & \cos\alpha \end{pmatrix}. \tag{2.3}$$

Using complex coordinates $z = x + iy = \sqrt{2j}e^{i\varphi}$, iz^* we obtain the simpler form

$$z' = e^{i\Omega(\frac{zz^*}{2})}z. \tag{2.4}$$

2.2. *Quasi-integrable maps*

We first consider a family of maps which are obtained as Poincaré sections of the Hamiltonian

$$H = \frac{p^2}{2} + F(x) + G(x)\sum_{n=1}^{\infty}\delta(t - nT). \tag{2.5}$$

Letting $f(x) = -F'(x)$, $g(x) = -G'(x)$, the equations of motion are

$$\dot{x} = p, \qquad \dot{p} = f(x) + g(x)\sum_{n=1}^{\infty}\delta(t - nT).$$

As a consequence, integrating in $[nT - \epsilon, nT + \epsilon]$ and letting $\epsilon \to 0$, we have

$$\begin{cases} x(nT+0) = x(nT-0), \\ p(nT+0) = p(nT-0) + g(x(nT-0)). \end{cases} \tag{2.6}$$

In order to obtain a mapping which relates x, p at $t = nT - 0$ to their values at $t = (n+1)T - 0$, we use the evolution from $t = nT + 0$ to $t = (n+1)T - 0$ determined by the integrable system $\dot{x} = p$, $\dot{p} = f(x)$. We write

$$\begin{pmatrix} x((n+1)T-0) \\ p((n+1)T-0) \end{pmatrix} = \mathcal{F}\begin{pmatrix} x(nT+0) \\ p(nT+0) \end{pmatrix}, \tag{2.7}$$

and substituting (2.6) into the r.h.s. of (2.7) after setting $x_n \equiv x(nT-0)$, $p_n \equiv p(nT-0)$, we obtain

$$\begin{pmatrix} x_{n+1} \\ p_{n+1} \end{pmatrix} = \mathcal{F}\circ\begin{pmatrix} x_n \\ p_n + g(x_n) \end{pmatrix}. \tag{2.8}$$

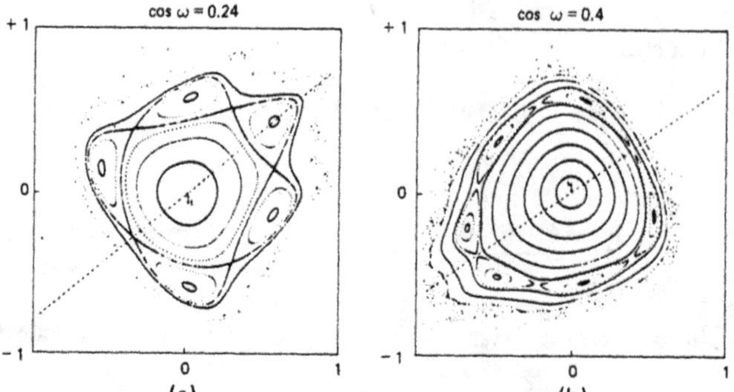

(a) (b)

Figure 10

Two well known examples are given by the *kicked pendulum* and *forced oscillators* which occur in modelling the beam-beam interactions of colliders. [17]

The Hamiltonian for the kicked pendulum is

$$H = \frac{j^2}{2} - \frac{\epsilon}{T}\cos\varphi \sum_{n=1}^{\infty} \delta(t - nT). \tag{2.9}$$

Observing that $g(\varphi) = -\frac{\epsilon}{T}\sin\varphi$ and $\mathcal{F}\binom{\varphi}{j} = \binom{\varphi + jT}{j}$, the map reads

$$\begin{cases} \varphi_{n+1} = \varphi_n + Tj_n - \epsilon \sin\varphi_n, \\ j_{n+1} = j_n - \frac{\epsilon}{T}\sin\varphi_n, \end{cases}$$

and, letting $r = Tj$, it takes the standard form first proposed in [21]

$$\begin{cases} r' = r - \epsilon\sin\varphi, \\ \varphi' = \varphi + r', \end{cases} \tag{2.10}$$

The Hamiltonian for the forced oscillator is

$$H = \frac{p^2}{2} + \frac{\omega^2}{2}x^2 + G(x)\sum_{n=1}^{\infty}\delta(t - nT). \tag{2.11}$$

In this case, we have $\mathcal{F}\binom{x}{p/\omega} = R(\omega t)\binom{x}{p/\omega}$ and, setting $y = p/\omega$, where $R(\alpha)$ is defined by (2.3), the map reads

$$\begin{pmatrix} x' \\ y' \end{pmatrix} = \begin{pmatrix} \cos\omega T & -\sin\omega T \\ \sin\omega T & \cos\omega T \end{pmatrix} \begin{pmatrix} x \\ y + \frac{g(x)}{\omega} \end{pmatrix}. \tag{2.12}$$

Using complex variables $z = x + iy$, it takes the simpler form

$$z' = e^{i\omega T}[z + \frac{i}{\omega}g(\frac{z + z^*}{2})] \, . \tag{2.13}$$

In Fig. 10, we show from [22] the orbits of the quadratic map corresponding to $g(x) = -x^2, T = 1$.

The perturbation parameter in this case is $|z|$, as one can easily verify scaling the variable according to $z \to \lambda z$. Close to the origin, slightly deformed invariant circles are observed; further away islands appear and a region of chaotic orbits. Very similar behavior is observed for the standard map (2.10).

2.3. Fixed points

The qualitative behavior of a map $\mathbf{x}' = F(\mathbf{x})$ is determined by its fixed points $\mathbf{x}_0$, that is, $F(\mathbf{x}_0) = \mathbf{x}_0$. If $\mathbf{x}_0$ is a fixed point of $F^n = F \circ F \circ .... \circ F$, that is, $F^n(\mathbf{x}_0) = \mathbf{x}_0$, then it is called a fixed point of order n or a n cycle.

Linearizing the map in the neighborhood of a fixed point,

$$\mathbf{x}' = \mathbf{x}_0 + F'(\mathbf{x}_0)(\mathbf{x} - \mathbf{x}_0) + O(|(\mathbf{x} - \mathbf{x}_0|^2), \tag{2.14}$$

the matrix $A = F'(\mathbf{x}_0)$ has unit determinant and according to the nature of its eigenvalues we have the following classification:

$$\begin{cases} \textit{hyperbolic} & \text{if } |\mathrm{Tr} A| > 2 \quad \text{real eigenvalues } \lambda \neq \pm 1, \\ \textit{elliptic} & \text{if } |\mathrm{Tr} A| < 2 \quad \text{complex eigenvalues } e^{\pm i\alpha}, \\ \textit{parabolic} & \text{if } |\mathrm{Tr} A| = 2 \quad \text{equal eigenvalues } \lambda = \pm 1, \end{cases}$$

where Tr denotes the trace. Using appropriate coordinates, the linear map (2.14) in the hyperbolic case reads

$$\begin{cases} x' = \lambda x, \\ y' = \lambda^{-1} y, \end{cases} \tag{2.15}$$

and in the elliptic case

$$\begin{pmatrix} x' \\ y' \end{pmatrix} = R(\alpha) \begin{pmatrix} x \\ y \end{pmatrix}. \tag{2.16}$$

Since xy and $x^2 + y^2$ are invariant, the respective trajectories are hyperbolas and ellipses.

The main question is whether the presence of neglected nonlinear terms preserves the topological structure of the orbits near the fixed points, so that we can find a new system of coordinates $(x, y) \to (\xi, \eta)$ such that the map takes a *normal form* which exhibits explicitly the invariant curves.

If the origin is an hyperbolic fixed point, the normal form reads

$$\begin{cases} \xi' = \Lambda(\xi\eta)\xi, \\ \eta' = \Lambda^{-1}(\xi\eta)\eta, \end{cases} \tag{2.17}$$

and if it is an elliptic point

$$\begin{pmatrix} \xi' \\ \eta' \end{pmatrix} = R(\Omega(\xi^2 + \eta^2)) \begin{pmatrix} \xi \\ \eta \end{pmatrix}, \tag{2.18}$$

or using complex coordinates $\varsigma = \xi + i\eta$ and ς^*,

$$\varsigma' = e^{i\Omega(\varsigma\varsigma^*)}\varsigma \tag{2.19}$$

and the complex conjugate equation that can be ignored.

2.4. The Poincaré-Birkhoff geometric theorem

We have seen that normal forms in the vicinity of an elliptic fixed point can be written if all the orbits are diffeomorphic to circles. The results on the geometric structure of orbits for Hamiltonian systems suggest that this cannot be expected in general and Moser's theorem [25] shows that the orbits of a perturbed twist map are diffeomorphic to circles only if the perturbation is sufficiently small and smooth and the rotation number diophantine.

A proof of nonintegrability of such maps goes back to Poincaré and we present it because it is simple and gives a deep insight on the breaking mechanism of invariant curves.

Theorem. Let T be a twist map and $r^* = 2\pi\frac{m}{n}$ a circle of periodic orbits. Let T_ϵ be a perturbed map

$$T_\epsilon : \begin{cases} \varphi' = \varphi + r + \epsilon f(r,\varphi), \\ r' = r + \epsilon g(r,\varphi), \end{cases} \tag{2.20}$$

with $f, g \in C^1(D_R), D_R = [0,R] \times \mathbf{T}$. Then there are two circles on which T_ϵ^n induces rotations in opposite directions and a curve Γ between them on which T_ϵ^n induces radial displacements. The intersections of Γ and $T_\epsilon^n\Gamma$ are fixed points (see Fig. 11), half being elliptic and half hyperbolic. The map T_ϵ has n as many elliptic and hyperbolic fixed points.

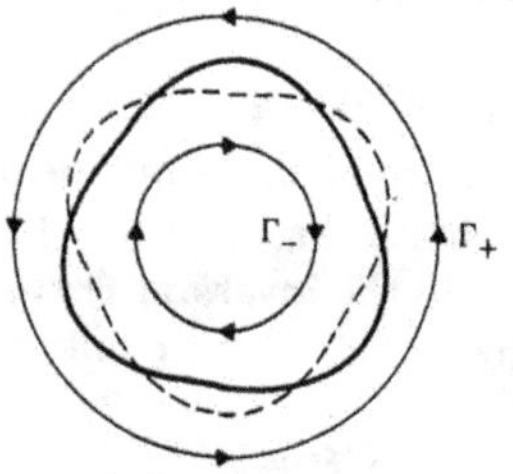

Figure 11

Sketch of proof: Since φ is a coordinate on the circle we can write

$$T_\epsilon^n : \begin{cases} \varphi' = \varphi + n(r - r^*) + \epsilon f_n(r,\varphi;\epsilon), \\ r' = r + \epsilon g_n(r,\varphi;\epsilon), \end{cases} \tag{2.21}$$

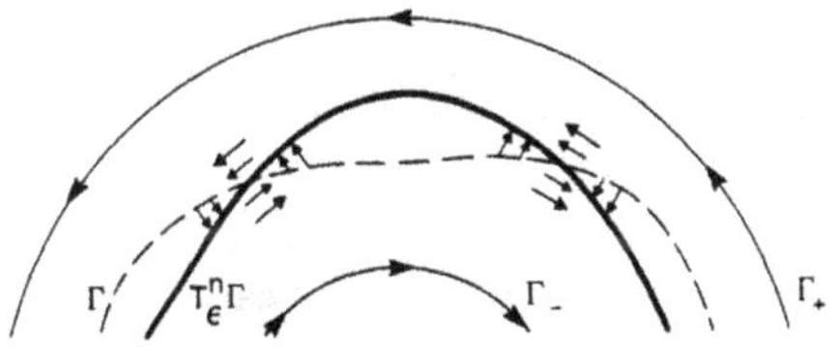

Figure 12

We suppose f and g to be continuous with their derivatives in D_R. Then in a smaller domain $D_{R-\delta}$ we can bound g_n and f_n, and letting A be a bound to the norms of f and g and their derivatives in D_R we have

$$\|f_n\|_{D_{R-\delta}} \leq \frac{n(n+1)}{2}A, \qquad \|\frac{\partial f_n}{\partial r}\|_{D_{R-\delta}} \leq An2^{n+1}e^{n\epsilon A}, \tag{2.22}$$

provided that $n\epsilon A \leq \delta$. Then choosing $\delta < \min\{\frac{n}{n+1}r^*, (R-r^*)\frac{n}{(2n+1)}\}$, the circles $r = r_\pm \equiv r^* \pm \frac{n\pm 1}{n}\delta$ are within $D_{R-\delta}$ and the rotations induced by T_ϵ^n are opposite since

$$\|\epsilon f_n\|_{D_{R-\delta}} \leq \frac{n+1}{2}\delta = \frac{n}{2}|r_\pm - r^*|. \tag{2.23}$$

If we impose the additional constraint $\delta < n2^{-n-3}$, which implies $\epsilon < 2^{-n-3}/A$, the twist condition is also satisfied by T_ϵ^n. Indeed, $\frac{\partial\varphi'}{\partial r} + n + \epsilon\frac{\partial f_n}{\partial r}$ does not change sign because from (2.22) we have

$$\|\frac{\epsilon}{n}\frac{\partial f_n}{\partial r}\|_{D_{R-\delta}} \leq 2^{n+1}\delta e^{\delta}/n \leq 1/2. \tag{2.24}$$

The curve Γ where the displacements are radial is defined by

$$r = r^* + \frac{\epsilon}{n}f_n(r,\varphi). \tag{2.25}$$

With the above conditions on δ the existence of a unique solution $r = r(\varphi)$ within the ring $r_- \leq r \leq r_+$ can be easily proved. Indeed, (2.23) insures the existence of a solution, while writing the first order Taylor expansion of the r.h.s. of (2.24) gives

$$|r - r^*| = \frac{\|\frac{\epsilon}{n}f_n\|_{D_{R-\delta}}}{1 - \|\frac{\epsilon}{n}\frac{\partial f_n}{\partial r}\|_{D_{R-\delta}}} \leq |r_\pm - r^*|. \tag{2.26}$$

Once the existence of the curve Γ where the displacements are radial is insured, then the area preserving condition implies that Γ and $T_\epsilon^n\Gamma$ must intersect. An inspection of the vector field in the neighborhood of the intersection points shows that elliptic and hyperbolic points alternate (see Fig. 12).

2.5. Instability

The Poincaré theorem explains the appearence of islands which surround the elliptic fixed points and are observed in any numerical computation. It also explains the appearence of chaotic regions which are associated to the hyperbolic fixed points. The behavior of a map near a hyperbolic fixed point is described by its normal form (2.15). The lines $\xi = 0$, $\eta = 0$ are called *stable* and *unstable manifolds* W_s, W_u if $\Lambda(0) = \lambda > 1$ (see Fig. 13).

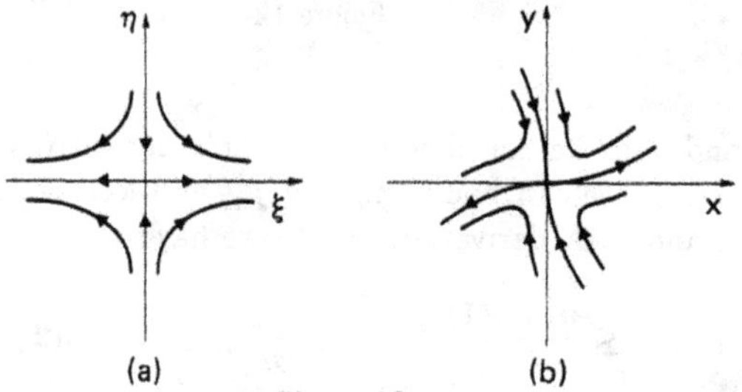

Figure 13

For integrable systems such as the pendulum, the stable und unstable manifolds join smoothly to form a *separatrix* (see Fig. 14).

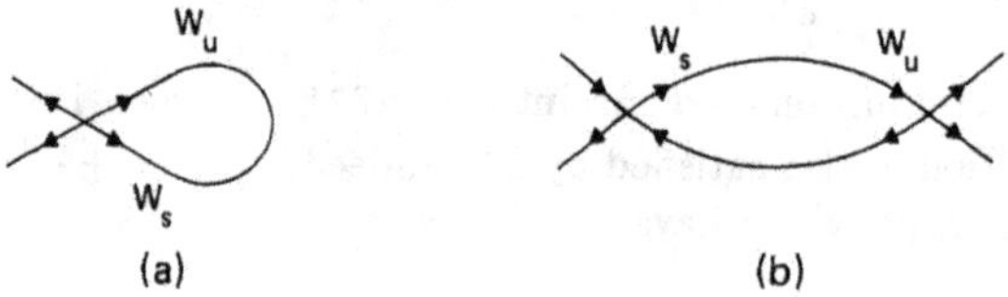

Figure 14

Generically this is not the case and W_s,W_u have a *transverse* intersection (the tangent vectors are distinct) called *homoclinic* point if W_s, W_u emanate from the same point and *heteroclinic* if they emanate from distinct points.

Letting $\mathbf{x}_0$,$\mathbf{y}_0$ be the first and second homoclinic intersections, we consider their images $\mathbf{x}_1, \mathbf{y}_1, \ldots, \mathbf{x}_k, \mathbf{y}_k \ldots$. Since the map preserves orientation, the sequences $\{\mathbf{x}_k\}$, $\{\mathbf{y}_k\}$ alternate and the loops on each side of W_s have equal areas (see Fig. 15).

Since the points $\mathbf{x}_k$, $\mathbf{y}_k$ become closer and closer the loops become exponentially longer and thinner. The loops of W_s can intersect the loops of W_u and an infinite hierarchy of intersections appears. As a consequence, the orbit fills densely a region of the plane and its Hausdorff dimension will be 2 rather than 1 as for the regular orbits (diffeomorphic to circles). The separation of two initially closed points is *exponential* while it is *linear* for regular orbits. For a more exhaustive exposition we refer to [16].

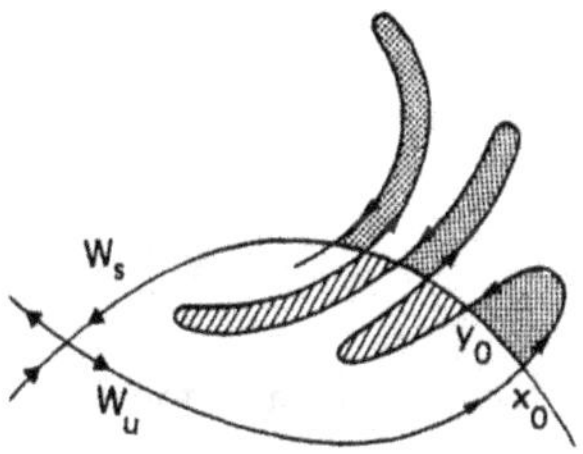

Figure 15

3. PERTURBATION THEORY FOR MAPS

3.1. Perturbation of nonresonant isochronous maps

We consider a map with an elliptic fixed point at the origin and letting $z = x + iy \in \mathbf{C}$ we write

$$z' = f(z, z^*) = e^{i\omega} z + \sum_{s \geq 2} f_s(z, z^*), \tag{3.1}$$

where f_s are homogeneous polynomials of degree s in z, z^* such that $\frac{\partial(z', z'^*)}{\partial(z, z^*)} = 1$. A sufficient condition is, for instance, that $f_s(z, z^*) = ie^{i\omega} g_s(z + z^*)$, where g_s is a polynomial with real coefficients. The perturbation parameter is $|z|$ as a simple scaling shows.

In the analysis carried out in this section, which is based on [23] and [26], a crucial role is played by the normal-form polynomials defined as follows. A polynomial $u(\varsigma, \varsigma^*)$ is defined to be in *normal form of order ℓ with respect to ω* if it satisfies the condition

$$\Delta_\ell u(\varsigma, \varsigma^*) \equiv u(e^{i\omega}\varsigma, e^{-i\omega}\varsigma^*) - e^{i\ell\omega} u(\varsigma, \varsigma^*) = 0. \tag{3.2}$$

The projector onto the subspace of the normal form polynomials of order ℓ will be denoted by Π_ℓ.

We first assume ω to be nonresonant, that is $\omega/2\pi$ irrational. The normal form polynomials of order ℓ are linear combinations of the monomials $(\varsigma\varsigma^*)^n \varsigma^\ell$ for $\forall n \in \mathbf{Z}$, and the aim of perturbation theory is to transform f into normal form of order 1 up to a finite order N. If the map is integrable, then it can be written in normal form in a neighborhood of the origin, while if it is nonintegrable

such a representation exists only locally. As a consequence, the remainder of the perturbation series cannot be expected to converge to 0 as $N \to \infty$.
Let

$$z = \Phi(\varsigma, \varsigma^*) = \varsigma + \phi(\varsigma, \varsigma^*) = \varsigma + \sum_{s=2}^{N} \phi_s(\varsigma, \varsigma^*) \tag{3.3}$$

be a transformation $\varsigma \to z$, where ϕ_s are homogeneous polynomials of order s and u be a map in normal form,

$$\varsigma' = u(\varsigma, \varsigma^*) = e^{i\Omega(\varsigma\varsigma^*)}\varsigma, \tag{3.4}$$

where

$$\Omega = \omega + \sum_{s=1}^{\left[\frac{N-1}{2}\right]} (\varsigma\varsigma^*)^s \Omega_{2s}. \tag{3.5}$$

We require the diagram

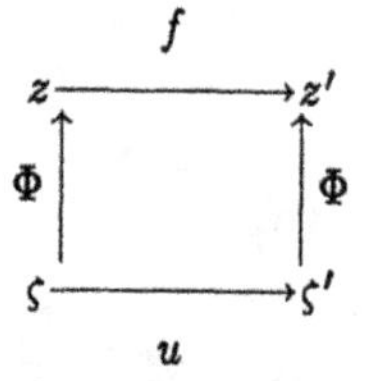

to commute up to order $|z|^{N+1}$, that is

$$f \circ \Phi = \Phi \circ u + O(|z|^{N+1}), \tag{3.6}$$

in order to determine Φ and u. We impose the additional condition on the transformation $\varsigma \to z$ that it be area preserving,

$$\tau(\varsigma, \varsigma^*) \equiv \frac{\partial(z, z^*)}{\partial(\varsigma, \varsigma^*)} = \frac{\partial(\Phi, \Phi^*)}{\partial(\varsigma, \varsigma^*)} = 1 + O(|\varsigma|^{N+1}). \tag{3.7}$$

Equation (3.6) explicitly reads

$$\phi(e^{i\Omega(\varsigma\varsigma^*)}\varsigma, e^{-i\Omega(\varsigma\varsigma^*)}\varsigma^*) - e^{i\omega}\phi(\varsigma, \varsigma^*) + \left[e^{i\Omega(\varsigma\varsigma^*} - e^{i\omega}\right]\varsigma =$$

$$= \sum_{s=2}^{N} f_s(\varsigma + \phi(\varsigma, \varsigma^*), \varsigma^* + \phi^*(\varsigma, \varsigma^*)) + O(|\varsigma|^{N+1}). \tag{3.8}$$

Replacing ϕ and Ω by their expansions (3.3), (3.5), the following recurrence relations are found:

$$\begin{cases} \Delta_1 \phi_{2n}(\varsigma, \varsigma^*) = Q_{2n}(\varsigma, \varsigma^*), \\ \Delta_1 \phi_{2n+1}(\varsigma, \varsigma^*) + i e^{i\omega} \Omega_{2n} \varsigma^{n+1} \varsigma^{*n} = Q_{2n+1}(\varsigma, \varsigma^*), \end{cases} \tag{3.9}$$

where Q_k depends on previously determined quantities. Letting

$$\phi_s(\varsigma,\varsigma^*) = \sum_{k=0}^{s} a_k^s \varsigma^k (\varsigma^*)^{s-k}, \tag{3.10}$$

the following result holds.

Theorem. If $\omega/2\pi$ is irrational the recurrence relations (3.9) determine ϕ_{2n} completely at even orders, Ω_{2n} and ϕ_{2n+1} except the coefficient a_{n+1}^{2n+1} of $\varsigma^{n+1}\varsigma^{*n}$ at odd orders.

Proof: Letting q_k^s be the coefficients of $\varsigma^k(\varsigma^*)^{s-k}$ for $Q_s(\varsigma,\varsigma^*)$, we have for even orders

$$\left[e^{2i\omega(k-n)} - e^{i\omega}\right] a_k^{2n} = q_k^{2n},$$

and the a_k^{2n} are all defined because the square braket never vanishes. For odd orders, we have

$$\left[e^{2i\omega(k-n-1)} - 1\right] a_k^{2n+1} - i\Omega_{2n}\delta_{k,n+1} = e^{-i\omega} q_k^{2n+1},$$

and since the square braket vanishes for $k = n+1$ it is evident that ω_{2n} is determined but a_{n+1}^{2n+1} remains arbitrary. In order to fix a_{n+1}^{2n+1}, we use the area preserving condition (3.7). However, this condition seems to impose constraints on the previously determined coefficients. In order to show that it fixes only the real part of a_{n+1}^{2n+1}, we need an additional result.

Theorem. If the coefficients τ_m^{2m} of $(\varsigma\varsigma^*)^m$ with $1 \le m \le [\frac{N}{2}]$ in the expansion of the Jacobian τ vanish, then $\tau = 1 + O(|\varsigma|^{N+1})$.

Proof: We first observe that

$$\tau(\varsigma,\varsigma^*) = \tau(e^{i\Omega}\varsigma, e^{-i\Omega}\varsigma^*), \tag{3.11}$$

because

$$\frac{\partial(z', z'^*)}{\partial(\varsigma,\varsigma^*)} = \frac{\partial(z', z'^*)}{\partial(z, z^*)}\,\frac{\partial(z, z^*)}{\partial(\varsigma,\varsigma^*)} = \frac{\partial(z', z'^*)}{\partial(\varsigma',\varsigma'^*)}\,\frac{\partial(\varsigma',\varsigma'^*)}{\partial(\varsigma,\varsigma^*)}$$

and because two Jacobians are unity since the maps f and u are area preserving. The result will be proved by induction: we assume it is true at order $n-1$, that is, $\tau = 1 + \tau_n + \ldots$, and from (3.11) we obtain at order n

$$\tau_n(\varsigma,\varsigma^*) = \tau_n(e^{i\omega}\varsigma, e^{-i\omega}\varsigma^*), \tag{3.12}$$

since $e^{i\Omega} = e^{i\omega} + O(|\varsigma|^2)$. Equation (3.12) becomes

$$\sum_{k=o}^{n} \left[e^{i\omega(2k-n)} - 1 \right] \tau_k^n \varsigma^k \varsigma^{*n-k} = 0.$$

The square braket vanishes only if $n = 2m$ is even and $k = m$; as a consequence, when such a condition is not met, $\tau_k^n = 0$. However, since τ_m^{2m} is zero by hypothesis, it follows that $\tau_n(\varsigma, \varsigma^*)$ vanishes $\forall n \leq N$.

We finally observe that τ can be written as

$$\tau(\varsigma, \varsigma^*) = \left[\frac{\partial \phi}{\partial \varsigma} + \frac{\partial \phi^*}{\partial \varsigma^*} \right] + \left[\frac{\partial \phi}{\partial \varsigma} \frac{\partial \phi^*}{\partial \varsigma^*} - \frac{\partial \phi^*}{\partial \varsigma} \frac{\partial \phi}{\partial \varsigma^*} \right], \tag{3.13}$$

and equating to zero the coefficient of $(\varsigma\varsigma^*)^n$, we obtain the real part of a_{n+1}^{2n} as a function of coefficients a_k^m with $m < 2n$.

The imaginary part of a_{n+1}^{2n} is arbitrary and is usually choosen equal to zero. However, it affects only the tansformation ϕ, not the rotation function Ω. Finally, we note that the transformation $\varsigma \to z$ can be inverted, so that

$$\varsigma = \psi(z, z^*) = z + \sum_{n=2}^{N} \psi_n(z, z^*) + O(|z|^{N+1}). \tag{3.14}$$

The first integral of motion is $|\psi(z, z^*)|$ and it defines the family of invariant curves in the original system of coordinates.

3.2. Perturbation of resonant isochronous maps

When ω is resonant, that is, $\omega/2\pi = p/q$, $p, q \in \mathbf{Z}$, the previous reduction to integrable form fails at orders $n \geq q - 1$. In this case, the normal form with respect to the resonant frequency ω must be used. We still look for a transformation $\varsigma \to z$ defined by (3.3) such that

$$\varsigma' = u(\varsigma, \varsigma^*), \tag{3.15}$$

where u is a polynomial of order N in normal form of order 1, that is,

$$\Delta_1 u(\varsigma, \varsigma^*) = 0.$$

Such a polynomial is the sum of monomials $\varsigma^r \varsigma^{*s}$, where $r - s = \pm lq + 1$ with $l \geq 0$, so that $r = s + lq + 1$ or $s = r + lq - 1$ (with $r \geq 1$ for $l = 0$). The monomials contributing to a normal form of order 1 can be written as $(\varsigma\varsigma^*)^m \varsigma^{lq} \varsigma$ and $(\varsigma\varsigma^*)^m \varsigma^{*lq} \varsigma$ for $m \geq 0, l \geq 0$ and ς^{*lq-1} for $l \geq 1$. When $q \to \infty$ only the monomials with $l = 0$, that is, $(\varsigma\varsigma^*)^m \varsigma$, survive as we found already in the nonresonant case.

The commutativity up to order N of the diagram leads to the equation

$$\phi(u\varsigma, u^*\varsigma^*) - e^{i\omega}\phi(\varsigma,\varsigma^*) + [u - e^{i\omega}\varsigma] = \\ = \sum_{s=2}^{N} f_s(\varsigma + \phi(\varsigma,\varsigma^*), \varsigma^* + \phi(\varsigma,\varsigma^*)) + O(|\varsigma|^{N+1}).$$

which to first order determines u according to $u = e^{i\omega}\varsigma + O(|\varsigma|^2)$ and to order n gives

$$\Delta_1\phi_n(\varsigma,\varsigma^*) + u_n(\varsigma,\varsigma^*) = Q_n(\varsigma,\varsigma^*),$$

where Q_n is known in terms of quantities determined at lower orders and Δ_1 is defined by (3.2). We observe that $\Pi_1\Delta_1\phi_n = \Delta_1\Pi_1\phi_n = 0$ and that $\Pi_1 u_n = u_n$. As a consequence, by projecting into the subspace of the normal form polynomials of order 1 and its complement we obtain

$$u_n = \Pi_1 Q_n$$

and

$$\Delta_1(1-\Pi_1)\phi_n = (1-\Pi_1)Q_n.$$

The polynomial u_n is determined, while the projection $\Pi_1\phi_n$ remains completly arbitrary; some conditions on it are imposed by the requirement that $z = \phi(\varsigma,\varsigma^*)$ is an area preserving transformation. Indeed, recalling that Π_0 denotes the projector onto the subspace of the normal-form polynomials of order 0, the following result holds.

Theorem. If the projection $\Pi_0\tau(\varsigma,\varsigma^*)$ of the Jacobian vanishes identically up to order N, then $\tau(\varsigma,\varsigma^*) = 1 + O\left(|\varsigma|^{N+1}\right)$ and

$$\sigma(\varsigma,\varsigma^*) \equiv \frac{\partial(u,u^*)}{\partial(\varsigma,\varsigma^*)} = 1 + O(|\varsigma|^{N+1}). \tag{3.16}$$

Proof: Equation (3.11) now reads

$$\tau(\varsigma,\varsigma^*) = \tau(u,u^*)\sigma(\varsigma,\varsigma^*). \tag{3.17}$$

Since u is a normal form of order 1, σ turns out to be a normal form of order 0. Indeed,

$$\sigma(e^{i\omega}\varsigma,\, e^{-i\omega}\varsigma) = \frac{\partial(e^{i\omega}u,\, e^{-i\omega}u^*)}{\partial(e^{i\omega}\varsigma,\, e^{-i\omega}\varsigma^*)} = \sigma(\varsigma,\varsigma^*).$$

This result is proved by induction: it is true for $n = 1$ and we assume it to be true at order $n-1$, so that

$$\begin{aligned} \tau(\varsigma,\varsigma^*) &= 1 + \tau_n(\varsigma,\varsigma^*) + \tau_{n+1}(\varsigma,\varsigma^*) + \ldots, \\ \sigma(\varsigma,\varsigma^*) &= 1 + \sigma_n(\varsigma,\varsigma^*) + \sigma_{n+1}(\varsigma,\varsigma^*) + \ldots. \end{aligned} \tag{3.18}$$

At order n Eq. (3.17) becomes, in view of (3.18) and $u = e^{i\omega}\varsigma + O(|\varsigma|^2)$,

$$\Delta_0 \tau_n(\varsigma, \varsigma^*) + \sigma_n(\varsigma, \varsigma^*) = 0, \tag{3.19}$$

where Δ_0 is defined by (3.2). We observe that $\Pi_0 \Delta_0 \tau_n = \Delta_0 \Pi_0 \tau_n \equiv 0$, so that projecting Eq. (3.19) with Π_0 we obtain $\Pi_0 \sigma_n = \sigma_n = 0$. Projecting with $1 - \Pi_0$ Eq. (3.19) gives $(1 - \Pi_0)\tau_n = 0$. As a consequence, only the normal form projection $\Pi_0 \tau_n$ remains arbitrary, and vanishes by hypothesis if $n \leq N$. As a consequence, $\sigma_n(\varsigma, \varsigma^*) = \tau_n(\varsigma, \varsigma^*) = 0$.

The condition $\Pi_0 \tau_0 = 0$ imposes a constraint on $\Pi_1 \phi_{n+1}$, which was arbitrary. We first observe that the expressions within the square brakets are equal to their complex conjugates, since $\frac{\partial \phi^*}{\partial \varsigma^*} = \left(\frac{\partial \phi}{\partial \varsigma}\right)^*$, $\frac{\partial \phi}{\partial \varsigma^*} = \left(\frac{\partial \phi^*}{\partial \varsigma}\right)^*$. From Eq. (3.13) we obtain

$$\tau_n(\varsigma, \varsigma^*) = \frac{\partial \phi_{n+1}}{\partial \varsigma} + \frac{\partial \phi^*_{n+1}}{\partial \varsigma^*} + S_n(\varsigma, \varsigma^*),$$

where S_n is a polynomial of order n which depends on $\phi_2, \ldots, \phi_n$ only. Thus,

$$\Pi_0 \tau_n = \frac{\partial}{\partial \varsigma} \Pi_1 \phi_{n+1} + \frac{\partial}{\partial \varsigma^*} \Pi_1 \phi^*_{n+1} + \Pi_0 S_n = 0.$$

$\Pi_0 \tau_n = 0$ gives the required constraints on $\Pi_1 \phi_{n+1}$.

3.3. The interpolation problem

For nonresonant maps, the normal form representation also allows one to obtain the first integral of motion $|\psi(z, z^*)|$ (see (3.14)), which gives a one-parameter family of invariant curves. It also allows one to write instantly the expression of an interpolating flow and the corresponding interpolating Hamiltonian.

In the resonant case, no explicit first integral of motion is obtained, but the interpolating flow is computed by a nontrivial construction. The associated interpolating Hamiltonian is time dependent but becomes time independent in a uniformly rotating frame and provides the desired first integral of motion, defining the invariant curves which are no longer diffeomorphic to circles.

We first consider a nonresonant map in the coordinate system where it is in normal form up to order N. The n-th iterate of the map for the initial point ς reads

$$\varsigma_n = e^{in\Omega(\varsigma\varsigma^*)}\varsigma, \qquad n \in \mathbf{Z},$$

and the interpolation is given by the flow Ψ:

$$Z \equiv \Psi(\varsigma, \varsigma^*; t) = e^{it\Omega(\varsigma\varsigma^*)}\varsigma, \qquad t \in \mathbf{R}. \tag{3.20}$$

Hamilton's equations of motion for the canonical coordinates Z, iZ^* are

$$\dot{Z} = -i\frac{\partial H}{\partial Z^*}, \qquad \dot{Z}^* = i\frac{\partial H}{\partial Z}, \tag{3.21}$$

where $H(Z,Z^*) = H^*(Z,Z^*)$ is the Hamiltonian. For the special flow (3.20) we have

$$H = h(ZZ^*), \qquad h(\rho) = -\int_0^\rho \Omega(r)dr. \tag{3.22}$$

Of Eqs. (3.21), only the first one needs to be considered, the second being its complex conjugate because H is real. The Hamiltonian function is a first integral for the orbit of the map and $H(\varsigma_n, \varsigma_n^*)$ =const gives the invariant curves which for the nonresonant case are circles.

The interpolation problem can be solved also in the resonant case, where the linear part of the map is $e^{i\omega}\varsigma$, with $\omega = 2\pi p/q$. We denote the flow with $Z = \Psi(\varsigma,\varsigma^*;t)$ and the corresponding vector field with $V(Z,Z^*;t)$, so that the equations of motion read

$$\dot{Z} = V(Z,Z^*;t), \qquad \dot{Z}^* = V^*(Z,Z^*;t), \tag{3.23}$$

the initial condition being $Z(0) \equiv \Psi(\varsigma,\varsigma^*;0) = \varsigma$. The solution is obtained within a perturbative scheme. We first expand V and Ψ in a series of homogeneous polynomials up to degree N:

$$V(Z,Z^*;t) = i\omega Z + V_2(Z,Z^*;t) + .. + V_n(Z,Z^*;t) + .., \tag{3.24}$$

and

$$Z \equiv \Psi(\varsigma,\varsigma^*;t) = e^{i\omega t}\varsigma + \Psi_2(\varsigma,\varsigma^*;t) + ..\Psi_n(\varsigma,\varsigma^*;t) + \dots \tag{3.25}$$

The initial conditions and the interpolation conditions at $t = 1$ read

$$\Psi_n(\varsigma,\varsigma^*;0) = \varsigma\delta_{n,1}, \qquad \Psi_n(\varsigma,\varsigma^*;1) = u_n(\varsigma,\varsigma^*). \tag{3.26}$$

We shall prove later that the interpolation condition at any integer time $t = n$ is also satisfied,

$$\Psi(\varsigma,\varsigma^*;n) = \varsigma_n \equiv u^n(\varsigma,\varsigma^*),$$

where $u^n = u \circ u \circ \cdots \circ u$ n times. If we substitute (3.25) into (3.24) and expand we find

$$\begin{aligned} V(Z,Z^*;t) =& i\omega e^{i\omega t}\varsigma + i\omega\Psi_2(\varsigma,\varsigma^*;t) + V_2(e^{i\omega t}\varsigma, e^{-i\omega t}\varsigma^*;t) \\ &+ \dots + i\omega\Psi_n(\varsigma,\varsigma^*;t) + S_n(\varsigma,\varsigma^*;t) + V_n(e^{i\omega t}\varsigma, e^{-i\omega t}\varsigma^*;t) + \dots, \end{aligned} \tag{3.27}$$

where S_n depends on Ψ_k and V_k with $k \le n-1$.

Theorem. If we choose

$$V_n(\varsigma,\varsigma^*;t) = e^{i\omega t}W_n(e^{-i\omega t}\varsigma, e^{i\omega t}\varsigma^*), \tag{3.28}$$

then the polynomials Ψ_n and W_n are both normal forms of order 1 ($\Pi_1\Psi_n = \Psi_n, \Pi_1 W_n = W_n$). The vector field V and the flux Ψ are then uniquely determined.

Proof: The result is evident for $n=1$ and proceeding by induction we suppose the result true at order $n-1$. We first show that $\Pi_1 S_n = S_n$. Indeed, letting P_n be the projector onto the subspace of homogeneous polynomials of order n, we have

$$S_n(e^{i\omega}\varsigma, e^{-i\omega}\varsigma^*;t) = P_n \sum_{k=c}^{n-1} V_k\Big(\sum_{j=1}^{n-1}\Psi(e^{i\omega}\varsigma, e^{-i\omega}\varsigma^*;t), \sum_{j=1}^{n-1}\Psi_j^*(e^{i\omega}\varsigma, e^{-i\omega}\varsigma^*;t)\Big) =$$
$$= e^{i\omega}S_n(\varsigma,\varsigma^*;t),$$

since $V_k(e^{i\omega}\varsigma, e^{-i\omega}\varsigma^*;t) = e^{i\omega}V_k(\varsigma,\varsigma^*;t)$ by (3.28) and the hypothesis that W_k is in normal form for $k \le n-1$, and $\Psi_k(e^{i\omega}\varsigma, e^{-i\omega}\varsigma^*) = e^{i\omega}\Psi_k(\varsigma,\varsigma^*)$ for $j \le n-1$. Taking into account (3.27), the equation of motion (3.23) becomes

$$\frac{\partial}{\partial t}\Psi_n(\varsigma,\varsigma^*;t) = i\omega\Psi_n(\varsigma,\varsigma^*;t) + V_n(e^{i\omega t}\varsigma, e^{-i\omega t}\varsigma^*;t) + S_n(\varsigma,\varsigma^*;t).$$

Multiplying by $e^{-i\omega t}$ and using (3.28) we integrate in the time interval $[0,t]$ and obtain for $n \ge 2$:

$$e^{-i\omega t}\Psi_n(\varsigma,\varsigma^*;t) = t\,W_n(\varsigma,\varsigma^*) + \int_0^t e^{-i\omega t'}S_n(\varsigma,\varsigma^*;t)dt', \tag{3.29}$$

where (3.26) has been used. The unknown W_n is determined at $t=1$ recalling that $\Psi_n(\varsigma,\varsigma^*;1) = u_n(\varsigma,\varsigma^*)$, and we have

$$W_n(\varsigma,\varsigma^*) = e^{i\omega}u_n(\varsigma,\varsigma^*) - \int_0^1 e^{-i\omega t'}S_n(\varsigma,\varsigma^*;t)dt'. \tag{3.30}$$

It is then evident that W_n is in normal form and by (3.29) Ψ_n is also in normal form. In addition, we notice that V_n cannot be chosen time independent, since in the integral on $[0,1]$ of $e^{-i\omega t}V_n(e^{i\omega t}\varsigma, e^{-i\omega t}\varsigma^*)$ all the terms not equal to $(\varsigma\varsigma^*)^m\varsigma$ vanish (recall that V is in normal form) while they are present in the r.h.s. of (3.30).

We are now ready to introduce the interpolating Hamiltonian. We first observe that defining the vector field $W = W_2 + W_3 + \ldots + W_n + \ldots$, related to V by

$$V(Z,Z^*;t) = i\omega Z + e^{i\omega t}W(e^{-i\omega t}Z, e^{i\omega t}Z^*), \tag{3.31}$$

the associated flux is $z = \overline{\Psi}(\varsigma, \varsigma^*; t)$, where

$$Z = e^{i\omega t} z, \tag{3.32}$$

since

$$\frac{dz}{dt} = W(z, z^*). \tag{3.33}$$

The final results are proved following Moser's arguments. [26)]

Lemma. The interpolation condition at $t = 1$ implies the interpolation at any $t = n \in \mathbf{Z}$; the interpolating flux is area preserving and the vector field W Hamiltonian.

Proof: We first observe that the vector field W is autonomous and that the semi-group property and the interpolation condition $\overline{\Psi}(\varsigma, \varsigma^*; 1) = e^{-i\omega} u(\varsigma, \varsigma^*)$ at $t = 1$ imply

$$\overline{\Psi}(\varsigma, \varsigma^*; n) = \overline{\Psi}(\varsigma, \varsigma^*; 1) \circ \cdots \circ \overline{\Psi}(\varsigma, \varsigma^*; 1) =$$
$$[e^{-i\omega} u(\varsigma, \varsigma^*)] \circ \cdots \circ [e^{-i\omega} u(\varsigma, \varsigma^*)] = e^{-in\omega} u^n(\varsigma, \varsigma^*),$$

where $u^n = u \circ \cdots \circ u$ n times and the property $u^k(e^{-i\omega}\varsigma, e^{i\omega}\varsigma^*) = e^{-ik\omega} u^k(\varsigma, \varsigma^*)$, easy to prove by induction, has been used.

We define now

$$D(\varsigma, \varsigma^*; t) \equiv \frac{\partial(Z, Z^*)}{\partial(\varsigma, \varsigma^*)} = \frac{\partial(z, z^*)}{\partial(\varsigma, \varsigma^*)},$$

taking (3.32) into account, and set

$$\delta(z, z^*) = \frac{\partial W}{\partial z} + \frac{\partial W^*}{\partial z^*}.$$

We claim that $D = 1 + O(|\varsigma|^{N+1})$, $\delta = O(|\varsigma|^{N+1})$, corresponding to the area preserving property of $\overline{\Psi}$ and the Hamiltonian property of the vector field W. Proceeding per absurdum, we suppose that $\delta = \delta_n + \cdots$ and $D = 1 + D_n + \cdots$ for some $n \leq N$, where as usual the index n denotes the degree of the polynomial. But at order n the equation satisfied by D,

$$\frac{\partial D}{\partial t}(\varsigma, \varsigma^*; t) = \delta(z, z^*) D(\varsigma, \varsigma^*; t),$$

would become

$$\frac{\partial}{\partial t} D_n(\varsigma, \varsigma^*; t) = \delta_n(\varsigma, \varsigma^*).$$

This implies $D = 1 + t\delta_n(\varsigma, \varsigma^*) + \ldots$. But this result contradicts (3.16), that is, the area preserving property of the given map, since for any integer m one has

$$D(\varsigma, \varsigma^*; m) = \prod_{k=o}^{m-1} \sigma(u^k(\varsigma, \varsigma^*), u^{k^*}(\varsigma, \varsigma^*)) = 1 + O(|\varsigma|^{N+1}).$$

We shall denote by $h(z,z^*)$ the Hamiltonian for the flux

$$z = \overline{\Psi}(\varsigma,\varsigma^*;t) = e^{-i\omega t}\Psi(\varsigma,\varsigma^*;t). \tag{3.34}$$

It is defined by

$$idh(z,z^*) = W^*dz - W\,dz^*. \tag{3.35}$$

Accordingly, the Hamiltonian $H(Z,Z^*;t)$ for the flux $Z = \Psi(\varsigma,\varsigma^*;t)$ is given by

$$i\,dH(Z,Z^*) = V^*dZ - V dZ^* \tag{3.36}$$

and one finds that

$$H(Z,Z^*;t) = -\omega ZZ^* + h(e^{-i\omega t}Z, e^{i\omega t}Z^*). \tag{3.37}$$

The final result is the following: there is a time dependent interpolating Hamiltonian, which in a rotating frame (with angular velocity ω) becomes time independent.

3.4. Examples of resonant behavior

For a resonant frequency $\omega = 2\pi p/q$, the first term appearing in the normal form, besides the terms $(\varsigma\varsigma^*)^n\varsigma$ which are present also in the nonresonant case, is ς^{*q-1}. Therefore if we consider the map

$$z' = e^{i\omega}[z - i(z+z^*)^{q-1}], \tag{3.38}$$

then the corresponding normal form is given by

$$\varsigma' = e^{i\omega}(\varsigma - i\varsigma^{*q-1}) + O(|\varsigma|^q). \tag{3.39}$$

The interpolating vector field W is given by

$$W = -iz^{*q-1}$$

and the Hamiltonian h reads

$$h = \frac{z^q}{q} + \frac{z^{*q}}{q}. \tag{3.40}$$

Using polar canonical coordinates $z = \sqrt{2\rho}e^{i\vartheta}$, we have

$$h = \frac{2}{q}(2\rho)^{q/2}\cos q\vartheta,$$

and the level lines, which interpolate the orbits of the map, have asymptotes at $\vartheta = \frac{2\ell-1}{2q}\pi$ for $\ell = 1,\ldots,q$.

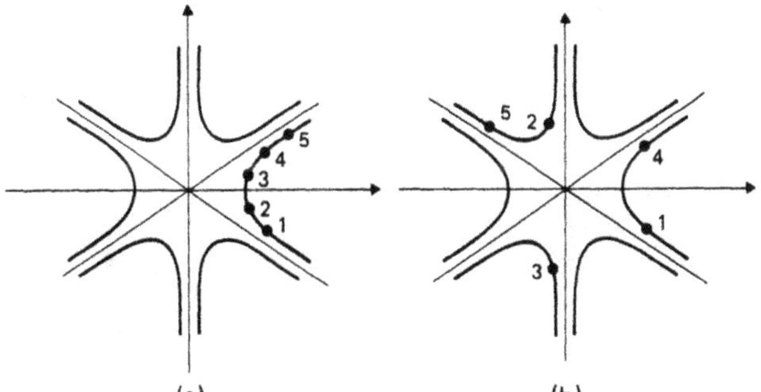

Figure 16

In Fig. 16a, we show for $q = 3$, $p = 1$ the level lines of h and an orbit at integer times $z(1), \ldots, z(5), \ldots$. In Fig. 16b, we show the orbit $Z(1), \ldots, Z(5), \ldots$ of H which is the true orbit of the map. Since $Z_k = e^{ik\omega} z_k$, there is each time a rotation by $\frac{2\pi}{3}$ and the point jumps from one branch to another (in this case the orbit of h has six symmetric branches).

Another interesting case is given by *almost resonant* frequencies $\omega = \omega_R + \epsilon$ where $\omega_R = 2\pi p/q$ and $\epsilon \ll 1$. The construction of the interpolating flow follows the scheme previously described for the genuine resonant case; the vector field V and the interpolating flow are given by (3.24) and (3.25), where $\omega = \omega_R + \epsilon$. The vector field V_n will still be given by (3.28), where $\omega = \omega_R$, namely,

$$V_n(\varsigma, \varsigma^*; t) = e^{i\omega_R t} W_n(e^{-i\omega_R t}\varsigma, e^{i\omega_R t}\varsigma^*),$$

and (3.29) is replaced by

$$\begin{aligned} e^{-i(\omega_R+\epsilon)t}\Psi_n(\varsigma, \varsigma^*; t) = & \int_0^t e^{-i\epsilon t'} W_n(e^{i\epsilon t'}\varsigma, e^{-i\epsilon t'}\varsigma^*)dt' + \\ & + \int_0^t e^{-i(\omega_R+\epsilon)t'} S_n(\varsigma, \varsigma^*; t')dt', \end{aligned} \tag{3.41}$$

which obviously reduces to (3.29) for $\epsilon = 0$. One still proves that S_n, W_n, Ψ_n are in normal forms with respect to ω_R. After determining W_n by evaluating the previous equation at $t = 1$, one defines the vector field $W_n(z, z^*) = i\epsilon z + W_2 + \ldots + W_n + \ldots$, related to V by

$$V(Z, Z^*; t) = i\omega_R Z + e^{i\omega_R t} W(e^{-i\omega_R t} Z, e^{i\omega_R t} Z^*),$$

so that the flux $z = \overline{\Psi}(\varsigma, \varsigma^*; t)$ of W is related to the flux $Z = \Psi(\varsigma, \varsigma^*; t)$ of V by

$$Z = e^{i\omega_R t} z.$$

Also in this case the interpolation at $t = 1$ implies the interpolation at any integer time and the vector field is Hamiltonian. The Hamiltonian h of the vector field W is in normal form with respect to ω_R and a constant of motion for the orbit of the map.

We first consider the map (3.38) with $\omega = \omega_R + \epsilon$, whose normal form with respect to ω_R is given by (3.39). It can be easily verified that $W_2 = \ldots = W_{q-2} = 0$ and that $S_{q-1} = 0$; as a consequence we obtain from (3.41):

$$W(z,z^*) = i\epsilon z + \frac{\epsilon q}{1 - e^{-i\epsilon q}} z^{*q-1} + O(|z|^q) = i\epsilon z - iz^{*q-1} + O(\epsilon|z|^{q-1}, |z|^q).$$

The corresponding interpolating Hamiltonian h reads

$$h = -\epsilon z z^* + \frac{z^q}{q} + \frac{z^{*q}}{q} + \ldots .$$

The Hamiltonian has an elliptic fixed point at the origin and q hyperbolic fixed points at $z_k = \epsilon^{1/(q-2)} e^{i\vartheta_k}$, where $\vartheta_k = \frac{2\pi k}{q}$ for $k = 1, \ldots, q$.
In Fig. 17 a sketch of the level lines of h for $q = 3$, $p = 1$ is shown.

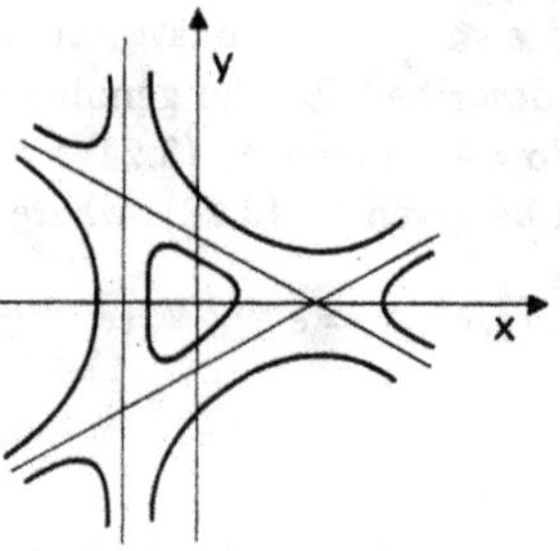

Figure 17

In the general case of a map having a quadratic perturbation and possibly higher order terms, if the frequency is very close to a resonant value ω_R, namely, $\omega = \omega_R + \epsilon$, the normal form reads

$$\varsigma' = e^{i(\omega_R+\epsilon)}\left[e^{i[\Omega_2 \varsigma\varsigma^* + \ldots + \Omega_{2m}(\varsigma\varsigma^*)^m]}\varsigma - i\alpha\varsigma^{*q-1}\right] + O(|\varsigma|^q),$$

where $m = \frac{q-2}{2}$ if q is even, $m = \frac{q-3}{2}$ if q is odd. The vector field W is given by

$$W(z,z^*) = i[\epsilon + \Omega_2 z z^* + \ldots + \Omega_{2m}(zz^*)^m]z + \alpha\frac{\epsilon q}{1 - e^{-i\epsilon q}} z^{*q-1} + O(|z|^q)$$

and the interpolating Hamiltonian is

$$h(z,z^*) = -\epsilon zz^* - \frac{\Omega_2}{2}(zz^*)^2 - \ldots - \frac{\Omega_{2m}}{m+1}(zz^*)^{m+1} - \frac{\alpha}{q}(z^q + z^{*q}) + O(\epsilon|z|^q, |z|^{q+1}).$$

Using polar canonical coordinates $z = \sqrt{2\rho}e^{i\vartheta}$, we write the Hamiltonian h as

$$h = \mathcal{H}(2\rho) - \frac{2\alpha}{q}(2\rho)^{q/2}\cos q\vartheta + \ldots, \tag{3.42}$$

where

$$\mathcal{H} = -\epsilon(2\rho) - \frac{\Omega_2}{2}(2\rho)^2 - \ldots - \frac{\Omega_{2m}}{m+1}(2\rho)^{m+1}.$$

The Hamiltonian h has an elliptic fixed point at the origin and assuming $\epsilon > 0$, $\Omega_2 < 0$, $\alpha > 0$ it has q elliptic fixed points at $z_{2k} = \sqrt{2\rho_+}e^{i\vartheta_{2k}}$, with $k = 1,\ldots,q$ and q hyperbolic fixed points at at $z_{2k+1} = \sqrt{2\rho_-}e^{i\vartheta_{2k+1}}$, with $k = 1,\ldots,q$, where $\vartheta_k = \frac{k}{q}\pi$ and $\rho_\pm$ are solutions of

$$\mathcal{H}'(2\rho_\pm) \mp \alpha(2\rho_\pm)^{q/2-1} = 0.$$

The solutions $\rho_\pm$ can be written as series in ϵ and one has

$$2\rho_\pm = -\frac{\epsilon}{\Omega_2} + O(\epsilon^2), \qquad \rho_+ - \rho_- = O(\epsilon^{q/2-1}).$$

We define an intermediate value ρ_0 by

$$\mathcal{H}'(2\rho_0) = 0$$

and observe that

$$2\rho_0 = -\frac{\epsilon}{\Omega_2} + O(\epsilon^2), \qquad \rho_\pm = \rho_0 + O(\epsilon^{q/2-1}).$$

We introduce a new radial coordinate ρ' to explore the orbits in a neighborhood of the fixed points

$$\rho = \rho_0 + \rho',$$

so that the Hamiltonian h becomes

$$h = \mathcal{H}(2\rho_0) + \frac{1}{2}(2\rho')^2\mathcal{H}''(2\rho_0)\left[1 + (2\rho')A_1 + \ldots + (2\rho')^{m-1}A_{m-1}\right] - \\ - \frac{2\alpha}{q}(2\rho_0)^{q/2}\cos q\vartheta\ (1 + \frac{\rho'}{\rho_0})^{q/2},$$

where

$$\mathcal{H}''(2\rho_0) = -\Omega_2[1 + O(\epsilon)], \qquad A_k = \frac{2}{k+2}\frac{\Omega_{k+2}}{\Omega_2}[1 + O(\epsilon)]\ \ k = 1,\ldots,m-1.$$

If the $|\Omega_k|$ do not increase too fast, we can approximate h with the Hamiltonian of a pendulum,

$$h = \mathcal{H}(2\rho_0) + \frac{\mathcal{H}''(2\rho_0)}{2}(2\rho')^2 - \frac{2\alpha}{q}(2\rho_0)^{q/2}\cos q\vartheta.$$

The separatrices have a maximal distance given by

$$\Delta = \left[\frac{2\alpha}{q\mathcal{H}''(2\rho_0)}\right](2\rho_0)^{q/4} \simeq \epsilon^{q/4}.$$

Since we are interested in the region within the separatrices we restrict ρ' to vary by $\epsilon^{q/4}$, and therefore for $q \geq 5$ we see that neglecting ρ' and ρ'/ρ_0 with respect to unity is perfectly justified for small ϵ. An example of such orbits is given for $q = 6$ by Fig. 7a.

The computed value of Δ gives an estimate of the width of the islands generated by the resonant frequency ω_R. We can use it also to estimate the area of the islands corresponding to the main resonances associated to a given irrational frequency. For this purpose we consider a diophantine frequency and more precisely we choose $\omega/2\pi$ to be a quadratic irrational. As it will be shown in the third paragraph of Section 3, denoting the leading rational approximations to such numbers by M_s/N_s, one has

$$\epsilon_s = \frac{\omega}{2\pi} - \frac{M_s}{N_s} \simeq \frac{1}{N_s^2}$$

and also $N_s = \beta e^{\alpha s}$ for s large. It will also be shown that for $2\rho \leq 2\rho_s = |\epsilon_s/\Omega_2|$ the Birkhoff series for Φ and Ω, truncated at $n \leq N_s$, have a convergent behavior; as a consequence the width Δ_s of the islands corresponding to the resonance M_s/N_s is given by

$$\Delta_s \sim \epsilon_s^{N_s/4} \sim N_s^{-N_s/2}.$$

Therefore, if we choose a disc of radius $R \sim \rho_{s_0} \sim N_{s_0}^{-2}$ it is not hard to show that the ratio of the area A_R of all the islands of the leading resonances within the disc (that is with $s \geq s_0$) to the area of the disc itself is bounded by

$$\frac{A_R}{\pi R^2} = \frac{1}{\pi R^2}\sum_{s\geq s_0} 2\pi\rho_s\Delta_s \simeq \frac{2}{R^2}\sum_{s\geq s_0}\frac{1}{N_s^2}N_s^{-N_s/2} \leq$$
$$\leq \frac{2}{N_{s_0}^2}\sum_{s\geq s_0} N_s^{-N_{s_0}/2} \simeq \frac{4}{\alpha R^2 N_{s_0}^3} N_{s_0}^{-N_{s_0}/2},$$

where the exponential approximation to N_s for s large was used and the sum was replaced by an integral over s. This ratio, whose behavior for R small is basically $\sqrt{R}^{1/2\sqrt{R}}$, is in agreement with Arnold's result, according to which the ratio of the measure of the invariant curves (that is the complement of the above set) to the measure of the disc approaches 1 as $R \to 0$.

3.5. Extensions to Hamiltonian maps of $\mathbf{R}^{2n}$

In various applications Hamiltonian maps of $\mathbf{R}^{2n}$ into itself are found and the reduction to normal form is a basic tool for understanding the behavior of the orbits in the neighborhood of an elliptic fixed point. The direct method used for maps of $\mathbf{R}^2$, based on a transformation, whose area preserving property was guaranteed a posteriori by determining recursively the coefficients left free by the the functional equation, is not applicable because the undetermined coefficients are overconstrained by the symplecticity condition and no consistent recursive scheme has been found to determine them. [27)]

I shall briefly describe a method developed by Bazzani,[27)] based on generating functions. We first recall that letting $\mathbf{x} = (x_1, \dots, x_n)$, $\mathbf{p} = (p_1, \dots, p_n)$ be the coordinates and momenta for a phase space $\mathbf{R}^{2n}$ and $H(\mathbf{x}, \mathbf{p})$ be a Hamiltonian, the transformation $(\mathbf{x}, \mathbf{p}) \to (\mathbf{z}, \mathbf{z}^*)$ with

$$\mathbf{z} = (z_1, \dots, z_n) \in \mathbf{C}^n, \qquad z_k = \frac{x_k + i p_k}{\sqrt{2}}$$

is canonical, since it preserves the Hamiltonian flow with respect to the new Hamiltonian $h(\mathbf{z}, \mathbf{z}^*) = -iH(\mathbf{x}, \mathbf{p})$. *

A canonical transformation $\Psi : (\mathbf{z}, \mathbf{z}^*) \to (\mathbf{Z}, \mathbf{Z}^*)$ where $\mathbf{Z} = \frac{\mathbf{X}+i\mathbf{P}}{\sqrt{2}}$ and $\mathbf{X}$, $\mathbf{P}$ are real conjugate variables, is obtained from a generating function $G(\mathbf{z}, \mathbf{Z}^*)$ and reads

$$\mathbf{z}^* = \frac{\partial G}{\partial \mathbf{z}}(\mathbf{z}, \mathbf{Z}^*) \quad , \quad \mathbf{Z} = \frac{\partial G}{\partial \mathbf{Z}^*}(\mathbf{z}, \mathbf{Z}^*). \tag{3.43}$$

Taking the complex conjugate of equation (3.43), we obtain

$$\mathbf{z} = \frac{\partial G^*}{\partial \mathbf{z}^*}(\mathbf{z}^*, \mathbf{Z}), \qquad \mathbf{Z}^* = \frac{\partial G^*}{\partial \mathbf{Z}}(\mathbf{z}^*, \mathbf{Z}), \tag{3.44}$$

and observe that G cannot be an arbitrary function of $\mathbf{z}$ and $\mathbf{Z}^*$, but must satisfy some functional constraints. If we considered a canonical transformation $(\mathbf{z}, \mathbf{w}) \to (\mathbf{Z}, \mathbf{W})$ in $\mathbf{C}^{2n}$, then the generating function $G(\mathbf{z}, \mathbf{W})$ would be completly free. In our case, we have the constraint $\mathbf{w} = \mathbf{z}^*$, $\mathbf{W} = \mathbf{Z}^*$ corresponding to the invariance of a $2n$ real dimensional subspace of $\mathbf{C}^{2n}$. For instance, if G is a polynomial with r terms, then $G(\mathbf{z}, \mathbf{Z}^*)$ will depend on r complex coefficients, but there will be r

---

* The Poisson brakets are not invariant since $[z_k, z_k^*] = -i$. This nonstandard choice is obtained with a scaling by i of the momentum from the standard variables $\mathbf{z}$, $i\mathbf{z}^*$ for which $[z_k, iz_k^*] = 1$ and the Hamiltonian is still H.

independent relations between them if it has to correspond to a real transformation $(\mathbf{q},\mathbf{p}) \to (\mathbf{Q},\mathbf{P})$ whose generating function is a polynomial with r real coefficients.

The functional equation which must be satisfied by $G(\mathbf{z},\mathbf{Z}^*)$ if is to correspond to a real tranformation is obtained by substituting (3.43) into (3.44) and reads

$$\begin{cases} \mathbf{z} = \dfrac{\partial G^*}{\partial \mathbf{z}^*}\left(\dfrac{\partial G}{\partial \mathbf{z}}(\mathbf{z},\mathbf{Z}^*), \dfrac{\partial G}{\partial \mathbf{Z}^*}(\mathbf{z},\mathbf{Z}^*)\right), \\ \mathbf{Z}^* = \dfrac{\partial G^*}{\partial \mathbf{Z}}\left(\dfrac{\partial G}{\partial \mathbf{Z}}(\mathbf{z},\mathbf{Z}^*), \dfrac{\partial G}{\partial \mathbf{Z}^*}(\mathbf{z},\mathbf{Z}^*)\right), \end{cases} \tag{3.45}$$

which must be identically satisfied. This implies that letting

$$G = \mathbf{z}\mathbf{Z}^* + G_3(\mathbf{z},\mathbf{Z}^*) + \ldots + G_s(\mathbf{z},\mathbf{Z}^*) + \ldots, \tag{3.46}$$

the following equation must be satisfied at order s :

$$\frac{\partial G_{s+1}(\mathbf{z},\mathbf{Z}^*)}{\partial \mathbf{Z}^*} + \frac{\partial G^*_{s+1}(\mathbf{Z}^*,\mathbf{z})}{\partial \mathbf{Z}^*} = Q_s(\mathbf{z},\mathbf{Z}^*), \tag{3.47}$$

where Q_s is a known polynomial of order s. As a consequence, a number of linear conditions equal to the number of coefficients coefficients of G_s is obtained.

An integrable Hamiltonian map of $\mathbf{R}^{2n}$ into itself is given by

$$\mathbf{Z}' = u(\mathbf{Z},\mathbf{Z}^*) = e^{i\Omega(\mathbf{Z}\mathbf{Z}^*)}\mathbf{Z}^*, \tag{3.48}$$

where

$$\Omega(\mathbf{Z}\mathbf{Z}^*) \equiv (\Omega_1(Z_1 Z^*{}_1,\ldots,Z_n Z^*{}_n),\ldots,\Omega_n(Z_1 Z^*{}_1,\ldots,Z_n Z^*{}_n))$$

with $\Omega_k(\rho_1,\ldots,\rho_n) = \frac{\partial H}{\partial \rho_k}(\rho_1,\ldots,\rho_n)$, H being a real function.

A generic map

$$\mathbf{z}' = f(\mathbf{z},\mathbf{z}^*) = e^{i\omega}(\mathbf{z} + f_2(\mathbf{z},\mathbf{z}^*) + \ldots),$$

where $\mathbf{z}' = e^{i\omega}\mathbf{z}$ means $z'_k = e^{i\omega_k} z_k$, $k = 1,\ldots,n$, can be conjugated to (3.48) with a canonical transformation $\Psi : (\mathbf{z},\mathbf{z}^*) \to (\mathbf{Z},\mathbf{Z}^*)$ generated by a function $F(\mathbf{z},\mathbf{Z}^*)$ if ω is *nonresonant*. Imposing the commutativity of the diagram

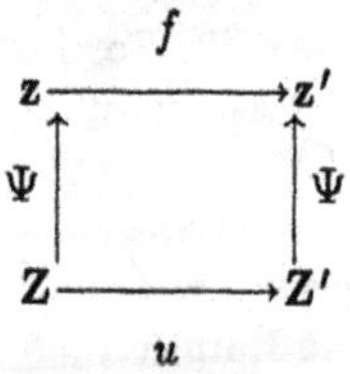

where $\mathbf{Z} = \Psi(\mathbf{z}, \mathbf{z}^*)$ is the trasformation generated by G, one obtains the desired functional equation,

$$\mathbf{z}^{*\prime} = \frac{\partial G}{\partial \mathbf{z}'}(\mathbf{z}', \mathbf{Z}^{*\prime}) = f^*(\mathbf{z}^*, \mathbf{z}).$$

Replacing $\mathbf{z}'$ and $\mathbf{Z}'$ by $f(\mathbf{z}, \mathbf{z}^*)$ and $u(\mathbf{Z}, \mathbf{Z}^*)$, respectively, and then $\mathbf{z}^*$ and $\mathbf{Z}$ by the r.h.s. of (3.43), one has the explicit expression of the functional equation for G and u:

$$\frac{\partial G}{\partial \mathbf{z}}(f(\mathbf{z}, \frac{\partial G}{\partial \mathbf{z}}(\mathbf{z}, \mathbf{Z}^*)), u^*(\mathbf{Z}^*, \frac{\partial G}{\partial \mathbf{Z}^*}(\mathbf{z}, \mathbf{Z}^*)) = f^*(\frac{\partial G}{\partial \mathbf{z}}(\mathbf{z}, \mathbf{Z}^*), \mathbf{z}). \tag{3.49}$$

The linear equations at order s are given by

$$e^{-i\omega}\left[\frac{\partial G_{s+1}}{\partial \mathbf{z}}(e^{i\omega}\mathbf{z}, e^{-i\omega}\mathbf{Z}^*) - \frac{\partial G_{s+1}}{\partial \mathbf{z}}(\mathbf{z}, \mathbf{Z}^*)\right] + u_s^*(\mathbf{Z}^*, \mathbf{z}) = P_s(\mathbf{z}, \mathbf{Z}^*), \tag{3.50}$$

where P_s are known polynomials.

The following theorem was proved by Bazzani: [27)]

Theorem. If f is a real Hamiltonian map and ω is nonresonant, then there is a real symplectic map with generating function G which satisfies conditions (3.45) and brings f into normal form. The indeterminacy in G corresponds to the abelian group of Birkhoff tranformations (which have the same form as for an integrable map).

When ω is resonant the conjugation is still possible with a map in normal form defined by

$$u(e^{i\omega}\mathbf{z}, e^{-i\omega}\mathbf{z}^*) = e^{i\omega}u(\mathbf{z}, \mathbf{z}^*).$$

The method is constructive and explicit algorithms can be written; the procedure for constructing the interpolating Hamiltonian can also be given.

4. NUMERICAL BEHAVIOR OF PERTURBATION SERIES

4.1. Numerical results for the Birkhoff series

We first describe the asymptotic properties of the Birkhoff series which bring into normal form the quadratic map

$$z' = f(z, z^*) \equiv e^{i\omega}[z - \frac{i}{4}(z + z^*)^2], \tag{4.1}$$

as they emerge from the numerical results obtained from the first $N = 100$ terms of the series. The rotation number $\omega/2\pi$ is the golden mean $\frac{\sqrt{5}-1}{2}$. Truncating the series at order N we have

$$f^m \circ \Phi - \Phi \circ u^m = O(|\varsigma|^{N+1}).$$

We define the discrepancy between the *actual* and *normal form* dynamics by

$$d = \frac{1}{m}|f^m \circ \Phi - \Phi \circ u^m|. \tag{4.2}$$

Letting $\varsigma = re^{i\vartheta}$, we observe that d does not depend appreciably on m and varies little with ϑ. The significant dependence is on N and r.

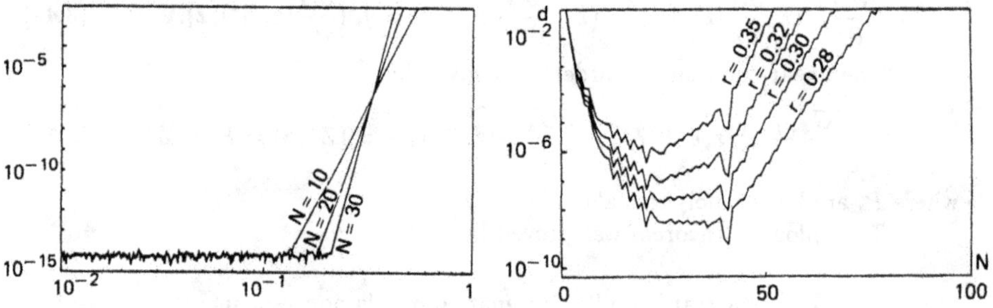

Figure 18

Fig. 18a shows $\log d(r, N)$ versus N for fixed r and the minima in N are clearly depicted. Fig. 18b shows $\log d$ versus $\log r$ for fixed N. Above the noise due to machine accuracy, we have straight lines with angular coefficient $N+1$, which are fitted with

$$d = c\left(\frac{r}{R}\right)^{N+1}. \tag{4.3}$$

However, as first observed in [28], the crossing point of the straight lines is the same only for values of N belonging to suitable intervals, where necessarily c and R are also constant. This means that $c(N)$ and $R(N)$ are piecewise constant functions of N,

$$\begin{cases} R = R_1, & c = c_1, \quad \overline{N}_1 < N < \overline{N}_2, \\ R = R_2, & c = c_2, \quad \overline{N}_2 < N < \overline{N}_3, \\ \cdots\cdots\cdots \end{cases} \tag{4.4}$$

The sequence $\overline{N}_s$ is increasing, while the sequence R_s is decreasing.

If we examine the series defining Φ and Ω, we also find a similar behavior, namely, that they are *piecewise geometric*,

$$\begin{cases} ||\Phi_n||_{L^2}^{1/2} / ||\Phi_{n+2}||_{L^2}^{1/2} \sim R_s, \\ |\Omega_{2n}|^{1/2} / |\Omega_{2n+2}|^{1/2} \sim R_s. \end{cases} \qquad \overline{N}_s < n < \overline{N}_{s+1},$$

The piecewise geometric behavior is even sharper if we consider the series defining the Fourier coefficients. Indeed, letting $\varsigma = re^{i\vartheta}$ from (3.3) and (3.10) we obtain:

$$\Phi = \sum_{n=1}^{N} r^n \sum_{k=0}^{n} a_k^n e^{i(2k-n)\vartheta} = \sum_{k=-N}^{N} \hat{a}_k(r) e^{ik\vartheta}, \tag{4.5}$$

where

$$\hat{a}_k(r) = \sum_{n=|k|}^{N} \hat{a}_k^n r^n, \qquad \hat{a}_k^n = \begin{cases} a^n_{\frac{n+k}{2}} & n+k \text{ even,} \\ 0 & n+k \text{ odd.} \end{cases} \tag{4.6}$$

In Fig. 19, the ratios $|\hat{a}_k^n|/|\hat{a}_k^{n+1}|$ are plotted for some fixed values of k and the same ratios $R_1, R_2, \ldots$ appear

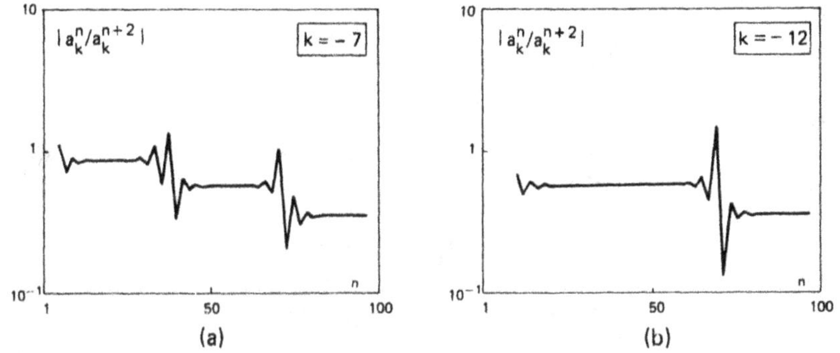

Figure 19

4.2. Resonance asymptotics

The behavior of the Birkhoff series observed in the numerical computations can be explained in terms of the contributions of the dominant resonances M_k/N_k which are obtained by truncating the continued fraction expansion of $\omega/2\pi$. Indeed, letting

$$\omega = 2\pi\frac{M_s}{N_s} + \epsilon_s, \tag{4.7}$$

we know that $|\epsilon_s/2\pi| \leq \frac{1}{N_s^2}$, while given any integer N for the rational M/N which best approximates $\omega/2\pi$, the estimate on the distance is only $|\omega/2\pi - M/N| \leq 1/N$.

For the quadratic irrationals (the entries of the continued fraction are all equal so that they are a countable set) explicit expressions for the N_s can be written and an exponential increase with s is found. The *radii of convergence* R_s turn out to be given by

$$R_s = |\frac{\epsilon}{\Omega_2}|^{1/2} \tag{4.8}$$

and can be easily understood, because this is exactly the value at which the first order approximation of $\Omega(r^2)$ takes the resonant value $2\pi M_s/N_s$. Indeed, letting

$\Omega(r) = \omega + \Omega_2 r^2$ for r small, the equation $\Omega(r) = 2\pi M_s/N_s$ is solved precisely for $r = R_s$.

A theoretical interpretation was given by considering the asymptotic behavior for $\epsilon_s \to 0$ that occurs when $s \to \infty$. We recall that the recurrence relation for the Fourier coefficients is

$$\hat{a}_k^n = \frac{\hat{q}_k^n}{e^{ik\omega} - e^{i\omega}}, \qquad n \geq |k|,$$

and $|e^{ik\omega} - e^{i\omega}| \simeq \epsilon_s$ for $k = \pm N_s + 1$. Therefore, the first coefficients $\simeq \epsilon_s^{-1}$ are $\hat{a}_{-N_s+1}^{N_s-1}$ and $\hat{a}_{N_s+1}^{N_s+1}$.

If we keep k fixed and increase the perturbation order, it can be shown that every two orders a new factor ϵ_s^{-1} appears, so that

$$|\hat{a}_k^{|k|+2m}| \simeq \epsilon_s^{-m}, \qquad k = \pm N_s + 1. \tag{4.9}$$

We recall that $\hat{a}_k^n$ with $n + k$ odd vanish.

By a direct asymptotic analysis it was proved [29,30] that as $s \to \infty$, so that $\epsilon_s \to 0$, the following relations hold

$$\begin{cases} a_{-N_s+1}^{N_s-1+2(m+1)} = -\dfrac{N_s + m + 1}{N_s + m}\dfrac{\Omega_2}{\epsilon_s} a_{-N_s+1}^{N_s-1+2m}, \\ a_{N_s+1}^{N_s+1+2(m+1)} = -\dfrac{m+2}{m+1}\dfrac{\Omega_2}{\epsilon_s} a_{N_s+1}^{N_s+1+2m}. \end{cases} \tag{4.10}$$

Due to the nonlinearity, the ϵ_s^{-1} contributions propagate to nearby Fourier coefficients according to the tree shown in the following table, where the powers of ϵ_s^{-1} are displayed.

	$-N_s-4$	$-N_s-3$	$-N_s-2$	$-N_s-1$	$-N_s$	$-N_s+1$	$-N_s+2$	$-N_s+3$	$-N_s+4$	$-N_s+5$	$-N_s+6$	$-N_s+7$
N_s-1						1						
N_s					1	·	1					
N_s+1				1	·	2	·	1				
N_s+2			2	·	2	·	2	·	1			
N_s+3		2	·	2	·	3	·	2	·	1		
N_s+4	2	·	3	·	3	·	3	·	2	·	1	
N_s+5	·	3	·	3	·	4	·	3	·	2	·	1

The propagation of the ϵ_s^{-1} contribution, which follows a triangular pattern, implies that every Fourier coefficient $\hat{a}_k(r)$ with $N_{s-1} < k < N_s$ will first be affected by an ϵ_s^{-1} contribution according to a geometric law and at some order $n > N_{s+1}$ by the ϵ_{s+1}^{-1} contributions (of the next leading resonance) which will prevail above the previous ones. All the next leading resonances will affect the same Fourier coefficient

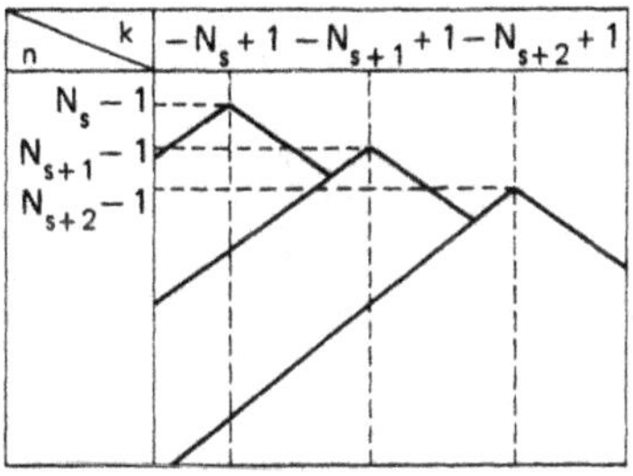

Figure 20

and the corresponding series diverge even though having an asymptotic property as numerically observed. In Fig. 20 the pattern of the resonance propagation is shown

We can see how the divergence shows up by plotting the images Γ of a circle $\varsigma = re^{i\vartheta}$, that is $z = \Phi(re^{i\vartheta}, re^{-i\vartheta})$, for different values of the truncation order N. We choose a disc of radius r such that $R_s < r < R_{s-1}$ (see Fig. 21a), and observe that:

For $N_{s-1} < n < N_s$, a *convergent* behavior is to be expected because the resonance $s-1$ dominates and we are within the convergence radius R_{s-1} associated to it. The curve Γ is indeed a slightly deformed circle (see Fig. 21b).

For $n > N_s$, the resonance s becomes dominant through its Fourier coefficients $\pm N_s + 1$ which grow by increasing n because we are outside the convergence radius associated to this resonance. The curve Γ becomes wiggly (see Fig. 21c). Its shape can be approximated by retaining only the $k = 1, \pm N_s + 1$ Fourier coefficients.

For $n \gg N_s$, the divergent behavior appears; the coefficients $k = \pm N_s + 1$ are so large that Γ first exhibits N_s cusps and then N_s loops of increasing size (see Fig. 21d). These loops are reminiscent of the N_s islands the true map exhibits as a resonance effect and show that the conjugation to the normal form is lost because Φ is no longer a homeomorphism.

In Fig. 22 we show the typical Fourier spectra for $N < N_s$ and for $N \gg N_s$.

It is easy to explain the pattern described above by considering the mapping

$$z = \varsigma + a\varsigma^{N_s+1}, \qquad a = Ae^{i\alpha},$$

which letting $\varsigma = re^{i\vartheta}$ has only the $k = 1,\ N_s + 1$ Fourier components. We set $z = Re^{i\Theta}$ and observe that the image curve Γ of the circle $|\varsigma| = r$ is homeomorphic to the circle only if

$$\frac{d\Theta}{d\vartheta} = \frac{r^2}{R^2}[1 + A^2 r^{2N_s}(N_s + 1) + Ar^{N_s}(N_s + 2)\cos(N_s\vartheta + \alpha)]$$

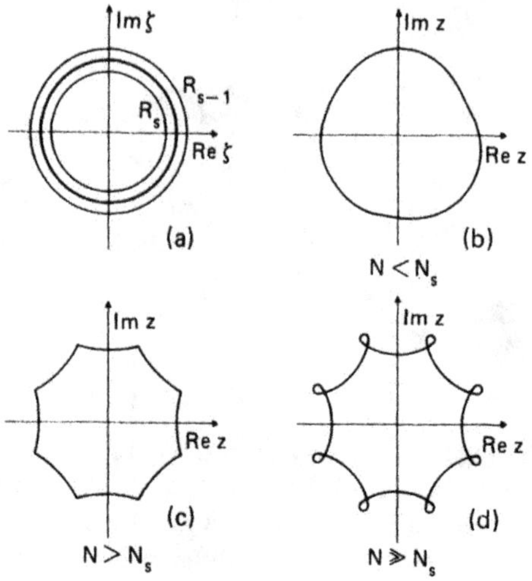

Figure 21

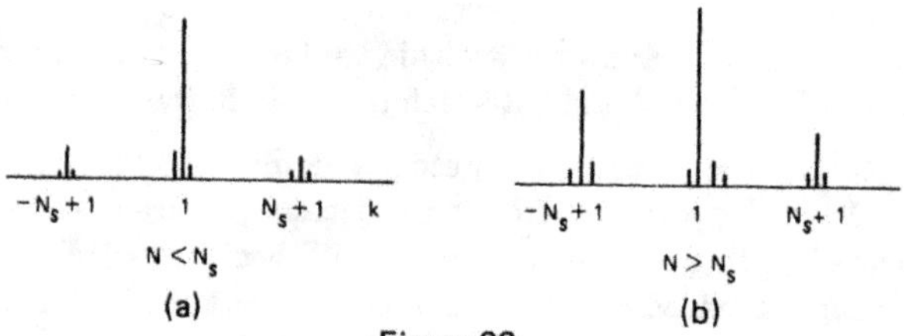

Figure 22

does not vanish. This implies that

$$r \leq r^* = [\frac{1}{A(N_s + 1)}]^{1/N_s}. \tag{4.11}$$

In the previous analysis when n is increased above N_s, A increases and consequently r^* decreases. Hence, keeping r fixed we have a value of n at which $\frac{d\Theta}{d\vartheta}$ vanishes and corresponds to the appearence of the N_s cusps. Above it, that is, for $r^* < r$, it is easy to show that Γ has N_s loops.

We finally notice that the order $\overline{N}_s$ at which the contribution of the resonance $2\pi M_s/N_s$ appears in the discrepancy funcion d or the Birkhoff series for Φ and Ω, decribed in Section 4.1, is higher than the resonance order ($\overline{N}_s > N_s$), because it takes some time, after it is switched on, in order to prevail on the previous resonance.

4.3. A solvable model problem.

We consider a series whose coefficients have ratios given by $|\epsilon_s/\Omega_2|^{1/2}$ and restrict the rotation numbers $\omega/2\pi$ to quadratic irrationals. The continued fraction expansion of these numbers has all the entries equal,

$$\frac{\omega}{2\pi} = p + \cfrac{1}{p + \cfrac{1}{p + \cfrac{1}{\cdots}}}, \quad p \in \mathbf{Z},$$

and it is easy to show that $\frac{\omega}{2\pi} = \frac{p+\sqrt{p^2+2}}{2}$. The approximations M_k/N_k obtained by truncating at order k satisfy the inequalities

$$\frac{p}{p^2+2}\frac{1}{N_s^2} \le \left|\frac{\epsilon_s}{2\pi}\right| \le \frac{1}{N_s^2}, \tag{4.12}$$

where ϵ_s is given by (4.7). One can also prove that asymptotically with s the N_s have an exponential behavior

$$N_s \sim \beta e^{\alpha s},$$

where $\alpha = \ln(\omega/2\pi)$ and $\frac{2}{p+\sqrt{4+p^2}} < \beta < \frac{1}{\sqrt{4+p^2}}$.

The model we propose is given by the series

$$F(\tau) = \sum_{n\ge 0} F_n(-\tau)^n, \tag{4.13}$$

where the coefficients F_n are defined by

$$F_n = F_{N_s}\left(\frac{1}{R_s}\right)^{n-N_s}, \qquad N_s + 1 \le n \le N_{s+1}, \tag{4.14}$$

so that the following recurrence relation holds for the F_{N_s}:

$$F_{N_s+1} = F_{N_s}\left(\frac{1}{R_s}\right)^{N_{s+1}-N_s}.$$

Since for the Birkhoff series the *convergence radii* R_s are given by $R_s = |\epsilon_s/\Omega_2|^{1/2}$ and for the quadratic irrationals the inequality (4.12) holds so that $c_1 N_s \le R_s^{-1} \le c_2 N_s$, our model will be close enough to a Birkhoff series if we choose

$$R_s^{-1} = cN_s$$

As a consequence, we have

$$\log F_{N_{s+1}} = \log F_{N_s} + (N_{s+1} - N_s)\log(cN_s) = \log F_{N_s} + \beta(e^{\alpha} - 1)[\log(c\beta) + \alpha s]e^{\alpha s}$$

and $\log F_{N_s+1}$ is given by a sum which for large s can be replaced by an integral which is easy to evaluate. Using Stirling's formula for the Γ function we can write

$$F_{N_s+1} \sim Ae^{BN_s+1}\Gamma\Big[\frac{1-e^{-\alpha}}{\alpha}N_{s+1}+1\Big]. \tag{4.15}$$

For N_s large enough, we can interpolate the coefficients with

$$F_n = Ae^{-Bn}\Gamma[Cn+1], \qquad C = \frac{1-e^{1-\alpha}}{\alpha}. \tag{4.16}$$

This new series is asymptotic to the function $f(r)$ defined by

$$f(r) = \int_0^\infty \frac{e^{-t}}{1+re^{B}t^{C}}dt$$

and the remainder of order N on the real positive axis is bounded by

$$|R_N(r)| \leq r^{N+1}e^{BN}\Gamma[C(N+1)+1].$$

For a fixed r, the remainder of the r.h.s. has a minimum for

$$N = \overline{N} = \frac{1}{C}e^{-(1+B/C)}r^{-1/C}$$

and we have

$$|R_{\overline{N}}| \leq e^{-\overline{N}C} = \exp[-r^{-1/C}e^{-(1+B/C)}]. \tag{4.17}$$

A model series for the conjugation function $\Phi(r,\theta) = \sum_{n\geq|k|} F_n(-r)^n$ is obtained by letting

$$a_k(r) = \sum_{n\geq|k|} F_n(-r)^n,$$

where the F_n are given by (4.14) and can be approximated asymptotically by (4.16). Then the series can be summed and one finds that it is asymptotic to a function and that the remainder is similar to (4.17), with different constants A and B.

4.4. The breakup of an invariant curve

The analysis of the Birkhoff series has given direct evidence for the nonglobal existence of invariant curves through the divergence of the conjugation function Φ. The origin of the divergence appears to be due to the accumulation of the contributions of the infinite sequence of leading resonances approaching the rotation number of the linear map. The asymptotic properties of the series allowed a nice interpolation of the existing invariant curves, in agreement with the existence of a C^∞ interpolation of the invariant tori proved in the case of Hamiltonian systems.[39]

Perturbation methods also allow one to investigate the behavior of a single invariant curve of fixed rotation number ω, when the perturbation strength λ is varied until it reaches the critical value $\lambda = \lambda_c$ at which the break-up occurs. In this case the series are found to be convergent, since the leading resonances give sporadic contributions (basically an ϵ^{-1} for each resonance rather than at each order as in the case of Birkhoff series).

We consider the standard map

$$S : \begin{cases} \varphi' = \varphi + r + \lambda \sin \varphi, \\ r' = r + \lambda \sin \varphi. \end{cases} \tag{4.18}$$

The imperturbed map has rotation number r. We conjugate a single invariant curve with rotation number ω to

$$T_\omega = \begin{cases} \omega' = \omega, \\ \Phi' = \Phi + \omega, \end{cases} \tag{4.19}$$

the transformation $(\varphi, r) \to (\Phi, \omega)$ being

$$\mathcal{F} : \begin{cases} r = \omega + \lambda v(\Phi, \omega; \lambda), \\ \varphi = \Phi + \lambda u(\Phi, \omega; \lambda), \end{cases} \tag{4.20}$$

where

$$u(\Phi + 2\pi) = u(\Phi), \qquad v(\Phi + 2\pi) = v(\Phi).$$

For ω fixed the transformation $\mathcal{F}$ satisfies the functional equation

$$S \circ \mathcal{F}(\omega, \Phi) = \mathcal{F} \circ T(\omega, \Phi). \tag{4.21}$$

In Fig. 23 we show the trajectories on $\mathbf{R} \times \mathbf{T}$.

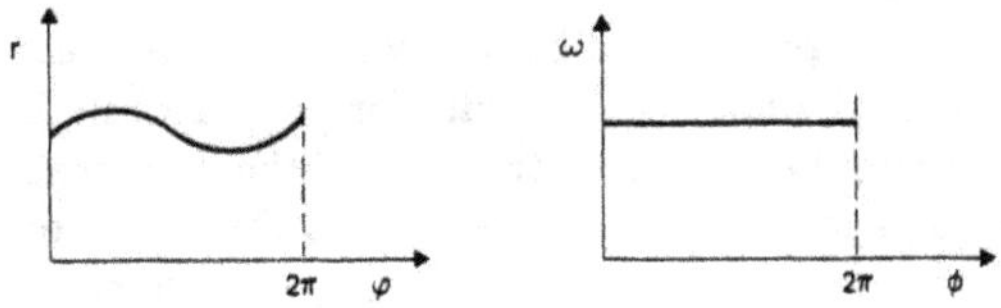

(a) (b)

Figure 23

If $\omega/2\pi$ is diophantine, then for $|\lambda|$ sufficiently small a unique solution $\mathcal{F}$ exists. We first write the functional equation explicitly:

$$\begin{cases} u(\Phi+\omega) - u(\Phi) = \lambda \sin(\Phi + u(\Phi)) + v(\Phi), \\ v(\Phi+\omega) - v(\Phi) = \lambda \sin(\Phi + u(\Phi)). \end{cases} \tag{4.22}$$

Substituting $v(\Phi)$ from the second equation into the first gives $u(\Phi+\omega) - u(\Phi) = v(\Phi+\omega)$, and after changing $\Phi+\omega$ into Φ we obtain a decoupled set of equations,

$$\begin{cases} u(\Phi+\omega) - 2u(\Phi) + u(\Phi-\omega) = \lambda \sin(\Phi + u(\Phi)), \\ v(\Phi) = u(\Phi) - u(\Phi-\omega). \end{cases} \tag{4.23}$$

Expanding u in a power series of λ,

$$u(\Phi) = \sum_{n\geq 1} \lambda^n u_n(\Phi),$$

yields

$$u_n(\Phi+\omega) - 2u_n(\Phi) + u_n(\Phi-\omega) = Q_n(\Phi),$$

where Q_n depends on $u_1, u_2, \ldots, u_{n-1}$, which are trigonometric polynomials

$$u_n(\Phi) = \sum_{k=-n}^{n} b_k^n e^{ik\Phi} \tag{4.24}$$

Finally, $u(\Phi)$ is written as a Fourier series

$$u(\Phi) = \sum_k b_k(\lambda) e^{ik\Phi}, \tag{4.25}$$

where

$$b_k(\lambda) = \sum_{n\geq|k|} b_k^n \lambda^n. \tag{4.26}$$

For a rotation number equal to the golden mean $\omega/2\pi = \frac{\sqrt{5}-1}{2}$, the perturbation series were computed up to the order $N = 140$ and for $|k| \leq N$ the sequences $|b_k^n|^{-1/n}$ $|k| \leq n \leq N$ were bounded below by 1, which seemed to be very close to their limit for $n \to \infty$. The Fourier spectrum for $|\lambda| \leq 1$ could be safely computed. This is at first surprising because the structure of the recurrence is very similar to the Birkhoff series. The leading resonances M_s/N_s first contribute at order $2N_s$ as ϵ_s^{-2}, but at higher orders the contributions ϵ_s^{-4}, $\epsilon_s^{-6}\ldots$ cancel exactly and we have

$$b_{2N_s}^{2N_s+2m} \sim \frac{1}{\epsilon_s^2}, \qquad m = 0, 1, 2, \ldots. \tag{4.27}$$

When $n = 2N_{s+1}$ a new contribution $\sim \epsilon_s^{-2}$ will appear (for the golden mean $N_{s+1} < 2N_s$ and the harmonics $2N_s, 3N_s, \ldots$ of N_s do not contribute).

The Fourier spectrum decreases exponentially:

$$b_k(\lambda) \simeq e^{-\beta(\lambda)|k|}. \tag{4.28}$$

The function $\beta(\lambda)$ is decreasing and vanishes for $\lambda = \lambda_c = 0.971$, a value in good agreement with previous estimates obtained by different methods. At the critical value $\lambda = \lambda_c$ a subsequence of Fourier coefficients decreasing linearly is found [35,36] (see Fig. 24b):

$$b_{N_s}(\lambda_c) = \frac{A}{N_s}$$

As a consequence, $u(\Phi, \omega; \lambda_c)$ is at most C^1 but the numerical results suggest that it is *not differentiable* and *self-similar* (see Fig. 24a). Good evidence is provided by renormalization group arguments, which show that at the breakup point the invariant curve becomes a fractal. [31,32,34] For $|\lambda| < \lambda_c$ the function $u(\Phi, \lambda)$ is analytic in Φ for $|\,\mathrm{Im}\Phi| < \beta(\lambda)$ and the image of the segment $0 \leq \mathrm{Re}\Phi \leq 2\pi$, $\mathrm{Im}\Phi = \beta(\lambda)$, which is the border of the analyticity strip, is also a fractal. [37,38]

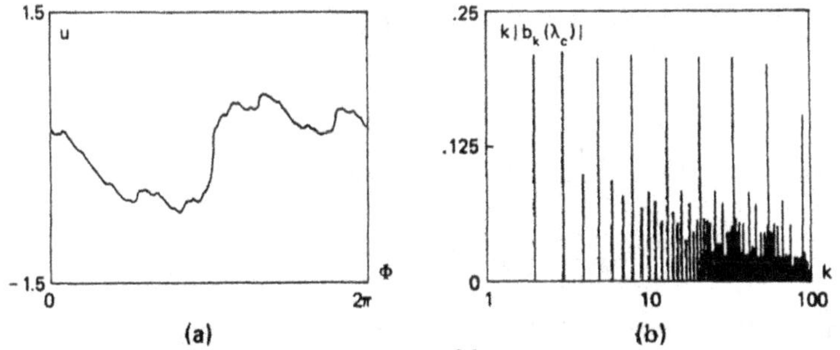

Figure 24

4.5. Another solvable model problem

As we did for the Birkhoff series, we propose a solvable model series which has the relevant features of the series defining the Fourier coefficients of the standard map

$$F(\lambda) = \sum \lambda^n F_n, \tag{4.29}$$

where

$$F_n = F_{N_s} \epsilon_s^{-2[\frac{n}{N_s}]}, \qquad N_s + 1 \leq n < N_{s+1}, \tag{4.30}$$

where the square brakets denote the integer part. The eventual contribution of harmonics is also taken into account if $N_{s+1} > 2N_s$. The recurrence relation for the F_{N_s} is

$$F_{N_{s+1}} = F_{N_s} N_s^{4\left[\frac{N_{s+1}}{N_s}\right]}.$$

Using $N_s \sim \beta e^{\alpha s}$, taking logarithms, and replacing the sum over s, defining $\log F_{N_s}$, by an integral, we obtain

$$F_{N_{s+1}} \simeq A e^{2\frac{[e^{\alpha}]}{\alpha}(\log N_{s+1})^2}, \tag{4.31}$$

which for large values of s can be interpolated by

$$F_n \sim A e^{2\frac{[e^{\alpha}]}{\alpha}(\log n)^2}. \tag{4.32}$$

The series is manifestly convergent with convergence radius 1.

The appropriate model for the Fourier coefficients is

$$u_k^n = \begin{cases} u_k^{N_s} \epsilon_s^{-2\left[\frac{n}{N_s}\right]}, & N_s + 1 \leq n \leq N_{s+1}, \quad n \geq |k|, \\ 0, & n < |k|. \end{cases}$$

Using similar approximation techniques, we find

$$u_k^n \simeq A e^{2\frac{[e^{\alpha}]}{\alpha}[(\log n)^2 - (\log |k|)^2]}$$

and for the sum $u_k(\lambda) = \sum_{n \geq |k|} \lambda^n u_k^n$ we obtain

$$u_k(\lambda) \simeq A e^{|k| \log \lambda - 2\frac{[e^{\alpha}]}{\alpha}(\log |k|)^2}, \tag{4.33}$$

in good qualitative agreement with the results observed for the standard map, that is, an exponential decrease for $\lambda < 1$.

5. THE SIEGEL PROBLEM

5.1. Perturbative solutions of the Siegel problem

A problem of small divisors which shares many features with the perturbation of an invariant curve of fixed rotation number is the Siegel center problem. [23,40] We consider a mapping of the complex plane C into itself,

$$z' = F(z) = e^{i\omega} z + f(z), \tag{5.1}$$

where $f(z)$ is analytic and $f(0) = f'(0) = 0$. The map is not area preserving, but in the linear case ($f = 0$) it has invariant circles with rotation number ω. It has been proved by Siegel that if $\omega/2\pi$ is diophantine, the nonlinear map 5.1 has invariant curves within a domain D (*Siegel domain*) and that they are analytically conjugated to circles within a disc of radius r_S (*Siegel radius*).

The transformation

$$z = \Psi(\varsigma) = \varsigma + \psi(\varsigma), \quad \psi(0) = \psi'(0) = 0, \tag{5.2}$$

which conjugates F with its linear part

$$\varsigma' = e^{i\omega}\varsigma \tag{5.3}$$

must satisfy the functional equation

$$\psi(e^{i\omega}\varsigma) - e^{i\omega}\psi(\varsigma) = f(\varsigma + \psi(\varsigma)). \tag{5.4}$$

The perturbative solution is easily obtained; for instance, if $f(z) = z^2$ we have

$$\psi(\varsigma) = \sum_{n=2}^{\infty} a_n \varsigma^n, \tag{5.5}$$

where the a_n are given by the recurrence relation

$$a_n = \frac{1}{e^{in\omega} - e^{i\omega}}[\delta_{n,2} + 2a_{n-1} + \sum_{k=2}^{n-2} a_k a_{n-k}],$$

the sum being absent for $n < 4$.

The asymptotic analysis of resonant contributions is easily made: letting $\frac{M_s}{N_s}$ be the order s leading resonance for the frequency ω (see (4.7)), we find that its first contribution of order $\sim \epsilon_s^{-1}$ appears in a_{N_s+1}. A new small divisor $\sim \epsilon_s$ is found only at order $n = 2N_s + 1$. Finally, at order $n = N_{s+1} + 1$, the contribution $\sim \epsilon_{s+1}^{-1}$ of the next resonance appears. The described frequency of occurrence of the divisors suggest that the series has a nonvanishing radius of convergence r_S. This radius can be computed by taking the limit of $|a_n|^{-1/n}$ or by mapping conformally the boundary of the Siegel domain, [43] which is the trajectory of a critical point, into a disc. [41,42] For a quadratic map with $f(z) = z^2$ and $\omega/2\pi$ equal to the golden mean, $r_S = 0.326$. Rigorous lower bounds to r_S are provided by the KAM type of estimates based on Newton's method.

5.2. Newton's method

The procedure we describe is standard but, the use of the following norms

$$||f||_r = \sum |f_n| r^n \tag{5.6}$$

for analytic functions

$$f(z) = \sum f_n z^n$$

and the choice of geometric convergence for the iterations, rather than superconvergence, yield explicit estimates in reasonable agreement with the numerical results. The scheme of the proof is reported because it is basically the same as for the Hamiltonian case (except for the ultraviolet cutoff that is not necessary in the Siegel problem). We first replace the functional equation with a linearized homological equation for $\Phi(\varsigma) = \varsigma + \varphi(\varsigma)$,

$$\varphi(a\varsigma) - a\varphi(\varsigma) = f(\varsigma), \qquad a = e^{i\omega}. \tag{5.7}$$

The mapping $F_1 = \Phi^{-1} \circ F \circ \Phi$ is no longer linear,

$$\varsigma' = F_1(\varsigma) = a\varsigma + f_1(\varsigma), \tag{5.8}$$

and the remainder $f_1(\varsigma)$ satisfies the functional equation

$$f_1(\varsigma) = Tf_1(\varsigma) \equiv \varphi(a\varsigma) - \varphi(a\varsigma + f_1(\varsigma)) + f(\varsigma + \varphi(\varsigma)) - f(\varsigma). \tag{5.9}$$

The aim of the method is to obtain a sequence of transformations $\Phi_n(\varsigma) = \varsigma + \varphi_n(\varsigma)$ and a sequence of mappings $F_n(\varsigma) = a\varsigma + f_n(\varsigma)$ analytic in discs of radius r_n, such that when $n \to \infty$ we have

$$r_n \to r_S, \quad ||f_n||_{r_n} \to 0, \quad \Phi_1 \circ \Phi_2 \circ \ldots \circ \Phi_n \to \Psi. \tag{5.10}$$

A diophantine condition on the frequency,

$$|a^n - 1|^{-1} \le \frac{1}{4}|n\frac{\omega}{2\pi} - m|^{-1} \le \gamma n^\mu, \qquad \mu \ge 1, \tag{5.11}$$

is crucial to guarantee the convergence of the solution of the linear homological equation

$$\varphi(\varsigma) = \sum_{n\ge 2} \frac{f_n}{a^n - a}\varsigma^n \tag{5.12}$$

and will be assumed. Indeed, assuming that $f(z)$ is analytic in a disc of $|z| \le R$, choosing for simplicity $\omega/2\pi$ to be a quadratic irrational (so that $\mu = 1$ in (5.11)), and defining

$$\delta(r) = ||f'(r)||_r = \sum_{n\ge 1} n|f_n|r^{n-1}, \tag{5.13}$$

the following estimate holds:

$$||\varphi||_r = \sum_n \frac{|f_n|}{|a^n - a|} r^n \le \sum_n |f_n|\gamma n r^n \le r\gamma \sum_n n|f_n|r^{n-1} = \gamma\delta r. \tag{5.14}$$

The estimate on the derivative of φ in a smaller disc of radius $r(1-\theta)$ is also easily obtained and reads

$$||\varphi'||_{r(1-\theta)} = \sum_{n\geq 2} |f_n| n r^{n-1}(1-\theta)^{n-1}(n-1)\gamma \leq$$

$$\leq \gamma \max_{n\geq 2}(n-1)(1-\theta)^{n-1} \sum_{n\geq 2} n|f_n| r^{n-1} \leq \frac{\gamma\delta}{e\theta}, \tag{5.15}$$

observing that the maximum of $n(1-\theta)^n$ for $n \geq 0$ is given by $\frac{-e^{-1}}{\log(1-\theta)} \leq (e\theta)^{-1}$.

The main theorem is proved by using the following technical lemma whose proof can be found in [43].

Lemma. If φ and ψ are analytic in a disc of radius ρ, $M = \max(||\varphi||_\rho, ||\psi||_\rho)$, and $f(z)$ is analytic in a disc of radius $\rho + M$, then

$$||f(\varsigma + \varphi) - f(\varsigma + \psi)||_\rho \leq ||f'||_{\rho+M} ||\varphi - \psi||_\rho. \tag{5.16}$$

Theorem. The remainder $f_1(\varsigma)$ is analytic in a disc of radius $r(1-\sigma-\eta)$, where

$$\sigma = \gamma\delta, \qquad \eta = \frac{\delta\sigma}{1-e^{-1}}, \tag{5.17}$$

and in a circle of radius $r_1 = r(1-\sigma-\eta-\hat{\sigma})$, where $\hat{\sigma}$ is an arbitrary real number such that $\sigma + \eta + \hat{\sigma} < 1$, the estimate on f_1' reads

$$||f_1'||_{r_1} \leq \delta_1 = \frac{\eta}{2\hat{\sigma}}. \tag{5.18}$$

Proof: The mapping T defined by $Th(\varsigma) = \varphi(a\varsigma) - \varphi(a\varsigma + h(\varsigma)) + f(\varsigma + \varphi(\varsigma)) - f(\varsigma)$ has a *unique fixed point* $h = f_1$ in the Banach space $\mathcal{B}$ of functions h such that

$$||h||_{r(1-\sigma-\eta)} \leq r\eta. \tag{5.19}$$

First one shows that T maps $\mathcal{B}$ into itself:

$$||Th||_{r(1-\sigma-\eta)} \leq ||\varphi'||_{r(1-\sigma-\eta)+||h||_{r(1-\sigma-\eta)}} ||h||_{r(1-\sigma-\eta)} +$$

$$+ ||f'||_{r(1-\sigma-\eta)+||\varphi||_{r(1-\sigma-\eta)}} ||\varphi||_{r(1-\sigma-\eta)}.$$

Using $||\varphi||_{r(1-\sigma)} \leq r\sigma$ and $||\varphi'||_{r(1-\sigma)} \leq e^{-1}$, obtained from (5.15) and (5.17), and using (5.13) and the definition of η, we obtain:

$$||Th||_{r(1-\sigma-\eta)} \leq \frac{r\eta}{e} + \delta r\sigma = r[\frac{\eta}{e} + \eta(1-\frac{1}{e})] = r\eta.$$

The contraction property also follows easily,

$$||Th_1 - Th_2||_{r(1-\sigma-\eta)} \le ||\varphi'||_{r(1-\sigma-\eta)+M}||h_1 - h_2||_{r(1-\sigma-\eta)} \le$$

$$\le ||\varphi'||_{r(1-\sigma)}||h_1 - h_2||_{r(1-\sigma-\eta)} \le \frac{1}{e}||h_1 - h_2||_{r(1-\sigma-\eta)},$$

where $M = \max(||h_1||_{r(1-\sigma-\eta)}, ||h_2||_{r(1-\sigma-\eta)}) \le r\eta$.

In order to evaluate f_1' we further restrict the radius, and letting $r_1 = r(1-\sigma-\eta-\hat{\sigma})$ and $x = \frac{\hat{\sigma}}{1-\sigma-\eta} < 1$ one has

$$||f_1'||_{r_1} = \sum_{n\ge2} n|f_{1n}|r^{n-1}(1-\sigma-\eta-\hat{\sigma})^{n-1} \le \sum_{n\ge2} n|f_{1n}|r^{n-1}(1-\sigma-\eta)^{n-1}(1-x)^{n-1} =$$

$$= \sum_{n\ge2} |f_{1n}|r^n(1-\sigma-\eta)^n \frac{1}{r\hat{\sigma}} \frac{n\hat{\sigma}}{(1-\sigma-\eta)}(1-x)^{n-1} \le \frac{\eta}{\hat{\sigma}} \max_{n\ge1} nx(1-x)^{n-1} \le \frac{\eta}{2\hat{\sigma}}.$$

5.3. The iteration scheme

The final result is obtained by iterating the one-step procedure just described. At step $n+1$ the remainder is f_{n+1}, analytic in a disc of radius r_{n+1}, where

$$||f_{n+1}'||_{r_{n+1}} \le \delta_{n+1}, \tag{5.20}$$

and the following relations hold:

$$r_{n+1} = r_n(1 - \sigma_n - \eta_n - \hat{\sigma}_n), \tag{5.21}$$

$$\delta_{n+1} = \frac{\eta_n}{2\hat{\sigma}_n}, \quad \eta_n = \frac{\delta_n \sigma_n}{1-e^{-1}}, \quad \sigma_n = \gamma\delta_n. \tag{5.22}$$

Usually a *superconvergent* iteration scheme $\delta_{n+1} = \delta_n^{1+\epsilon}$ with $\epsilon > 0$ is adopted. We prefer to choose *geometric* convergence by imposing

$$\delta_{n+1} = \chi\delta_n, \tag{5.23}$$

which implies

$$\sigma_{n+1} = \chi\sigma_n \quad \eta_{n+1} = \chi^2\eta_n \quad \hat{\sigma}_{n+1} = \chi\hat{\sigma}_n.$$

As a consequence, we can write

$$\delta_n = \chi^n\delta_0, \quad \sigma_n = \chi^n\sigma_0, \quad \eta_n = \chi^{2n}\eta_0, \quad \hat{\sigma}_n = \chi^n\hat{\sigma}_0, \tag{5.24}$$

where

$$\sigma_0 = \gamma\delta_0, \quad \eta_0 = \frac{\gamma\delta_0^2}{1-e^{-1}}, \quad \hat{\sigma}_0 = \frac{\eta_0}{2\delta_1} = \frac{\gamma\delta_0}{\chi(1-e^{-1})^2}. \tag{5.25}$$

The final resul reads:

Theorem. Letting $r_0 < R$ and $\delta_0 = \delta(r_0) = ||f'(r_0)||_{r_0}$, the map F is conjugated with its linear part, at least within a disc of radius r_∞ given by

$$r_\infty \geq r_0[1-\sigma_0-\eta_0-\hat{\sigma}_0]^{\frac{1}{1-\chi}}. \tag{5.26}$$

Proof: Indeed, using (5.21) and (5.24), we have

$$r_\infty = r_0\prod_{n=0}^{\infty}(1-\sigma_n-\eta_n-\hat{\sigma}_n) \geq r_0 \exp\sum_{n=0}^{\infty}\log[1-\chi^n(\sigma_0+\eta_0+\hat{\sigma}_0)] =$$

$$= r_0\exp\Big[-\sum_{n=0}^{\infty}\sum_{k=1}^{\infty}\frac{\chi^{nk}}{k}(\sigma_0+\eta_0+\hat{\sigma}_0)^k\Big] = r_0\exp\Big[-\sum_{k=1}^{\infty}\frac{(\sigma_0+\eta_0+\hat{\sigma}_0)^k}{k(1-\chi^k)}\Big].$$

A lower bound to r_∞ is obtained by replacing in the last sum χ^k with χ. The series within square brakets can then be summed and gives $\frac{1}{1-\chi}\log(1-\sigma_0-\eta_0-\hat{\sigma}_0)$ in agreement with (5.26).

The convergence of $\Psi_n = \Phi_1\circ\Phi_2\circ\ldots\circ\Phi_n$ to Ψ follows immediatly. Indeed, letting $z = \varsigma_0$ and

$$\begin{cases} \varsigma_0 = \Phi_1(\varsigma_1) = \varsigma_1 + \varphi_1(\varsigma_1), \\ \ldots \qquad \ldots \\ \varsigma_{n-1} = \Phi_n(\varsigma_n) = \varsigma_n + \varphi_n(\varsigma_n), \end{cases}$$

we can write

$$z = \Psi_n(\varsigma_n) = \varsigma_n + \psi(\varsigma_n) = \varsigma_n + \varphi_1(\varsigma_1) + \ldots + \varphi_n(\varsigma_n).$$

The function Φ_n is defined on a disc of radius r_n and its range is within a disc of radius r_{n-1}. Indeed,

$$||\Phi_n||_{r_n} \leq r_n + ||\varphi_n||_{r_n} \leq r_{n-1}(1-\sigma_{n-1}) + ||\varphi_n||_{r_{n-1}(1-\sigma_{n-1})} \leq r_{n-1}.$$

As a consequence,

$$||\psi_n||_{r_n} \leq \sum_{j=1}^{n}||\varphi_j||_{r_j} \leq \sum_{j=1}^{n}||\varphi_j||_{r_{j-1}(1-\sigma_{j-1})} \leq \sum_{j=1}^{n} r_{j-1}\sigma_{j-1} \leq \frac{r_0\sigma_0}{1-\chi}.$$

5.4. A numerical example.

We compute an explicit lower bound to the Siegel radius for the quadratic map $f(z) = z^2$ and a rotation number equal to the golden mean $\omega = 2\pi\frac{\sqrt{5}-1}{2}$. In this case $\delta(r) = 2r$ and $\mu = 1$, $\gamma = 0.5364\ldots$. We recall that

$$r_\infty \geq r\left[1 - 2\gamma r - 4\gamma\frac{r^2}{1-e^{-1}} - \frac{\gamma r}{\chi(1-e^{-1})}\right]^{\frac{1}{1-\chi}}$$

and that a maximization over r and χ should be carried out. Setting for simplicity $2\gamma \sim 1$ and $\chi = \frac{1}{2}$, we obtain

$$r_\infty \geq r\left[1 - r - \frac{2r^2}{1-e^{-1}} - \frac{r}{1-e^{-1}}\right]^2 = r[1 - 2.6r - 3.6r^2]^2.$$

If we restrict r to vary in the interval $[0, \frac{1}{9}]$, then

$$r_\infty \geq r[1 - 2.6r - \frac{3.6}{9}r]^2 = r[1-3r]^2.$$

The maximum occurs for $r = \frac{1}{9}$ and

$$r_\infty \geq \frac{1}{9}(1-\frac{1}{3})^2 = \frac{4}{81} = 0.049$$

to be compared with the numerical value $r_S = 0.326$ obtained from the conformal mapping of the trajectory of the critical point and the series analysis with the Hadamard criterion. The numerical evaluation of a finite number of steps with more accurate estimates for the norms and the introduction of perturbative information on f_1 and f_2 allows one to improve the estimate up to $r_\infty \geq 0.18$. Even better results were recently obtained by de la Llave by modifying the geometric setting. We conclude by observing that the superconvergent character of Newton's method shows up if we look at the Taylor series of f_n and φ_n. Indeed, if $f(z) = z^2 + \ldots$, then

$$f_n(z) = z^{2^n+1}[1 + O(z)], \qquad \varphi_{n+1}(z) = z^{2^n+1}[1 + O(z)],$$

as can be verified by induction on the functional equation for f_{n+1} which gives $f_{n+1} = f_n'(z)\varphi_n(z)[1 + O(z)]$.

Following a similar scheme, the extensions to the Siegel problem in $\mathbf{R}^n$ and in infinite dimensions were recently considered.[46,47)]

AKNOWLEDGMENTS

I wish to thank G. Benettin, S. Marmi, G. Servizi for useful discussions and A. Bazzani for his remarks on the normal form reduction and the construction of the interpolating Hamiltonians. I wish also to thank A.W. Sáenz for inviting me to deliver these lectures in Medellín and the local organizers for their warm hospitality.

REFERENCES.

Section 1

[1] Arnold, V.I. , *Méthodes Mathématiques de la Mécanique classique* (MIR, Moscow, 1976).

[2] Arnold, V.I. , *Chapitres supplémentaires à la théorie des équations différentielles* (MIR, Moscow, 1977).

[3] Arnold, V.I., and Avez, A. , *Ergodic problems of classical mechanics* (Benjamin, New York, 1968).

[4] Gallavotti, G. , *The elements of mechanics* (Springer, New York, 1983).

[5] Giacaglia, G.E.O. , *Perturbation methods in nonlinear systems,* Appl. Math. Sciences, Vol. 8 (Springer, 1972).

[6] Lichtenberg, A.J. and Lieberman, M.A., *Regular and stochastic motion* Appl. Math. Sciences, Vol. 38 (Springer, 1983).

[7] Kolmogorov, A.N. , Dokl. Akad. Nauk. SSSR **98,** 527 (1954)

[8] Arnold, V.I. , Russ. Math. Surv. **18,** 9 (1963)

[9] Nekhoroshev, N.N. , Russ. Math. Surv. **18,** 85 (1977)

[10] Giorgilli, A., and Galgani, L. , *"Rigorous estimates for the series expansions of Hamiltonian perturbation theory",* preprint (1986)

[11] Benettin, G. and Gallavotti, G. , J. Stat. Phys. **44,** 293 (1986)

[12] Melnikov, V. K. , Trans. Moscow Math. Soc. **12,** 1 (1963)

[13] Chirikov, B. V. , Phys. Reports **52,** 256 (1977)

[14] Poincaré, H. , *Les méthodes nouvelles de la mécanique céleste* (Gauthier-Villars, Paris, 1893), Vol. II, pp. 1,2.

Section 2

[15] Hénon, M. , *Comportement chaotique des systèmes déterministes,* Les Houches, 1981 (North Holland, Amsterdam, 1983), p. 53.

[16] Lanford, O., in:*Regular and chaotic motions in dynamic systems,* edited by G. Velo and A.S. Wightman (1985).

[17] Dragt, A. , *Nonlinear dynamics and the beam-beam interaction,* edited by M. Month and J.C. Herrera, AIP Conf. Proc. **57,** 143 (1979)

[18] Birkhoff, G.D. , *Dynamical systems,* Am. Math. Soc. Collog. Publ., Vol. IX (Am. Math. Soc., Providence, R.I., 1927).

[19] Birkhoff, G.D. , Acta Math. **43,** 1 (1920).

[20] Helleman, R.H.G. , in: *Fundamental problems in statistical mechanics* edited by E.G.D. Cohen (North Holland, Amsterdam, 1980).

[21] Greene, J.M. , J. Math. Phys. **20,** 1183 (1979).

[22] Hénon, M. , Quart. Appl. Math. **27,** 291 (1969).

Section 3

[23] Siegel, C.L., and Moser, J. , *Lectures in celestial mechanics* (Springer, Berlin, 1971).
[24] Moser, J. , *Stable and random motions in dynamical systems,* (Princeton Univ. Press, 1973).
[25] Moser, J., Nachr. Akad. Wiss. Göttingen Math. Phys. Kl. **1** 1 (1962).
[26] Moser, J. , *Lectures on Hamiltonian systems,* in: Memoirs Am. Math. Soc., Vol. 81 (Am. Math. Soc., Providence, R.I., 1968).
[27] Bazzani, A., *"Teoria perturbativa per mappe Hamiltoniane",* Tesi della Universitá di Padova (1987).

Section 4

[28] Roels,J. and Hénon, M. , Bull. Astr. Soc. **2**, 267 (1967).
[29] Servizi, G., Turchetti, G., and Benettin, G. , Phys. Lett. **95A,** 11 (1983).
[30] Servizi, G., and Turchetti, G. , Nuovo Cimento **95B,** 121 (1986).
[31] Shenker, S.J., and Kadanoff, L.P. , J. Stat. Phys. **27,** 631 (1982).
[32] McKay, R.S., Physica **7D,** 283 (1983) .
[33] Arnold, V.I., Trans. Am. Math. Soc. **46,** 213 (1965).
[34] Shenker, S.J., Physica **5D,** 405 (1982).
[35] Zanetti, G., and Turchetti, G., Lett. Nuovo Cimento **41,** 90 (1984) .
[36] Benettin, G., Turchetti, G., and Zanetti, G., Phys. Lett. **105A,** 436 (1984).
[37] Greene, J.M., and Percival, I.C., Physica **3D,** 530 (1981).
[38] Percival, I.C., Phisica **6D,** 67 (1982) .
[39] Chierchia, L., and Gallavotti, G. , Nuovo Cimento **67B,** 277 (1982) .

Section 5

[40] Siegel, C.L., Annals of Math. **43,** 607 (1942).
[41] Manton, N.S., and Nauenberg, M., Comm. Math. Phys. **89,** 555 (1983).
[42] Herman, M.R., Comm. Math. Phys. **99,** 593 (1985).
[43] Liverani, C., and Turchetti, G., J. Stat. Phys. **45,** 1071 (1986).
[44] De la Llave, M., J. Math. Phys. **24,** 2118 (1983).
[45] De la Llave, M., *"Proof of accurate bounds in small denominator problems",* preprint (1986).
[46] Marmi, S., Turchetti, G., and Vittot, M. , *"Estimates on the frequencies allowed by an infinite dimensional Siegel theorem",* preprint (1986).
[47] Marmi, S., *"Diophantine conditions and KAM estimates for the Siegel theorem in* $\mathbf{C}^N$*",* preprint (1986).

AN ELEMENTARY INTRODUCTION TO STOCHASTIC PROCESSES

Luis Vázquez
Departamento de Física Teórica
Facultad de Ciencias Físicas
Universidad Complutense
28040 - Madrid, Spain

CONTENTS

1. BASIC ELEMENTS OF STOCHASTIC PROCESSES

1.1. Introduction

It has long been known from microscopic observations that particles suspended in a liquid are in a state of constant, highly irregular motion. Such motion is called Brownian motion. The name derives from the English botanist Robert Brown, who noticed in 1827 the erratic movement of particles suspended in a fluid. A complete account of the history of the Brownian motion can be found in Nelson.[1)]

The main features of Brownian motion are the following:

(1) The motion is highly irregular.

(2) Two particles appear to move independently.

(3) The composition and density of the particles have no effect.

(4) The mobility of the suspended particles is larger when:

- The particles are smaller
- The fluid is less viscous
- The temperature is higher

(5) The motion is permanent, it never stops.

Such properties support the kinetic theory of the fluids, i.e., Brownian motion is caused by the impacts of the neighboring molecules of the fluid. On the other hand, property (5) indicates that Brownian motion is an equilibrium phenomenon.

1.2. Langevin equation for Brownian motion

The Langevin equation describes Brownian motion and is derived from the following ingredients:

A. A particle of mass m moving through a viscous medium experiences a friction force $\vec{F}_R$. For low velocities, Stokes'law is valid: $\vec{F}_R = -\alpha \vec{V}$ (For a sphere of radius R in a fluid of viscosity η, the damping coefficient is given by $\alpha = 6\pi R\eta$).

In the absence of additional forces the equation of motion for the particle is

$$m \frac{d\vec{V}}{dt} = -\alpha \vec{V}, \tag{1}$$

Whence

$$\vec{V}(t) = \vec{V}(0)\, e^{-\gamma t}.$$

$\tau_R = 1/\gamma = m/\alpha$ is the Relaxation time. We note that (1) is a deterministic differential equation, i.e., the velocity $\vec{V}(t)$ at time t is completely determined by its initial value.

B. The equipartition energy law of classical statistics

When we have a large number of particles, the total energy of the system is

$$E = \text{Kinetic Energy} \begin{pmatrix} \text{Translation} \\ + \\ \text{Rotation} \\ + \\ \text{Vibration} \end{pmatrix} + \text{Potential Energy of Vibration.}$$

In a first approximation this is accurate enough. In the framework of classical statistical mechanics, it is possible to prove that if the number of particles is large, if Newtonian mechanics holds, and if the system is in equilibrium at the absolute temperature T, then all the quadratic (independent) terms in the energy, have the same average value, and this average value depends only on the temperature and is

$$\langle E \rangle = \frac{1}{2} KT, \tag{2}$$

where k is Boltzmann's constant. This result was deduced by Maxwell. Such independent modes of energy absorption are called degrees of freedom.

C. Equation (1) is valid only if the mass of the particle is large. In this case the velocity due to thermal fluctuations is negligible. On the contrary, when the mass of the particle is very small, the thermal velocity may be observable:

$$\frac{1}{2} m \langle V^2 \rangle = \frac{1}{2} KT \quad \text{(one dimension)},$$

$$V_{th} = \langle V^2 \rangle^{1/2} = (KT/m)^{1/2}. \tag{3}$$

Thus for a small mass the deterministic Eq. (1) must be modified in order to obtain the correct thermal energy.

D. The modification of Eq. (1) consists in adding a fluctuating force such that its properties are given only on the average. The force is a stochastic or random force. For the sake of simplicity, let us consider the one-dimensional motion

$$m \frac{dV}{dt} = -\alpha V + F(t), \tag{4}$$

$$F(t) \Longrightarrow \frac{1}{2} m \langle V^2 \rangle = \frac{1}{2} KT. \tag{5}$$

In view of the nature of the fluctuating force, the following remarks are relevant:

To solve the coupled equations of motion associated to our system (molecules of the fluid and the particle) represents an impossible task because of the large number of molecules $\sim 10^{23}$. We must keep in mind that in such equations of motion no stochastic force is necessary. On the other hand, we do not know the initial values of all molecules of the fluid; thus the exact motion of the particle in the fluid cannot be calculated. In this way two systems (fluid + particle), identical except for the initial values of the fluid, define different motions of the particle.

This lack of information on the system suggests the necessity of considering an ensemble of systems: fluid + particle (Gibbs ensemble). And the force F(t) is different for each system, so that we have to consider averages of the force on the ensemble. Each system represents a realization of the stochastic force. We can visualize the force as a random sequence of impulses (positive and negative) which are homogeneous in time, unrelated to each other and to the velocity of the particle. The force represents the collisions of different molecules of the fluid with the immersed particle.

We have from Eq.(4)

$$\frac{dV}{dt} = -\gamma V + S(t) , \tag{6}$$

where S(t) = F(t)/m is the Langevin force. Equation (6) is the famous Langevin Equation, which is a linear stochastic differential equation.

Let us analyze the average properties of the Langevin force:

(a) $$\langle S(t) \rangle = 0 . \tag{7}$$

In effect, by considering averages in the Langevin equation we get

$$\frac{d}{dt} \langle V(t) \rangle = -\gamma \langle V(t) \rangle + \langle S(t) \rangle .$$

And $\langle S(t) \rangle = 0$, because $\langle V(t) \rangle \xrightarrow[t \to \infty]{} 0$ since the particle is "localized" in the fluid. From that we obtain

$$\langle V(t) \rangle = \langle V(0) \rangle e^{-\gamma t} .$$

(b) $$\langle S(t) S(t') \rangle = 2D \delta(t-t') . \tag{8}$$

This correlation function is determined "a posteriori" in order to satisfy the equipartition energy law. It turns out to be $D = \gamma KT/m$

According to the above observations about the nature of the Langevin force, we can assume $\langle S(t)\, S(t') \rangle = 0$ for $|t-t'| \geq \tau_o$, where τ_o is the duration time of a collision. This is related to the fact that the collisions of the particle with molecules of the fluid are approximately independent. Usually $\tau_o \ll \tau_R = 1/\gamma$, so the limit $\tau_o \to 0$ is a reasonable approximation.

1.3. Solution of the Langevin equation

According to the theory of stochastic differential equations,[2] the Langevin equation has a unique solution

$$V(t) = V(o)\, e^{-\gamma t} + e^{-\gamma t} \int_0^t S(t')\, e^{\gamma t'} dt' . \tag{9}$$

By using the properties (7) and (8) of the Langevin force we obtain[3]

(a) $$\langle V(t) \rangle = \langle V(o) \rangle\, e^{-\gamma t}, \tag{10}$$

the equation mentioned above.

(b) $$\langle V^2(t) \rangle = \langle V^2(o) \rangle\, e^{-2\gamma t} + \frac{D}{\gamma} . \tag{11}$$

For $t \gg 1/\gamma$ we have

$$\langle V^2(t) \rangle = \frac{D}{\gamma} = \frac{2}{m} \langle E \rangle = \frac{KT}{m},$$

in agreement with the equipartition energy law for stationary states.

Hence we obtain for the constant D the value

$$D = \gamma K T / m \,. \tag{12}$$

(c) The velocity correlation function is

$$\langle V(t_1)V(t_2)\rangle \begin{cases} = V^2(0)\, e^{-\gamma(t_1+t_2)} + \dfrac{D}{\gamma}\left(e^{-\gamma|t_1-t_2|} - e^{-\gamma|t_1+t_2|}\right), \\ \simeq \dfrac{D}{\gamma}\, e^{-\gamma|t_1-t_2|} \qquad \text{for} \qquad t_1, t_2 >> 1/\gamma \,. \end{cases} \tag{13}$$

For large times, the velocity correlation function is only a function of the time difference and is independent of the initial velocity.

(d) The mean-square value of the particle displacement is given by

$$\begin{aligned} \langle (X(t)-X_0)^2\rangle &= \int_0^t dt' \int_0^t dt'' \langle V(t')\, V(t'')\rangle \\ &= \left(V_0^2 - \frac{D}{\gamma}\right)\frac{(1-e^{-\gamma t})^2}{\gamma^2} + \frac{2D}{\gamma^2}\, t - \frac{2D}{\gamma^3}(1-e^{-\gamma t}), \end{aligned} \tag{14}$$

where X_o and V_o are the position and velocity of the particle at time $t = 0$.

If the initial velocity distribution corresponds to the stationary state, $\langle V_o^2\rangle = D/\gamma$, then the first term on the right-side of (14) vanishes.

For very large times, $t >> 1/\gamma$, the leading term in (14) is the linear one:

$$\langle (X(t)-X_0)^2\rangle \sim \frac{2D}{\gamma^2}\, t = \frac{2KT}{\gamma m}\, t \,. \tag{15}$$

This relation is the famous Einstein result for Brownian motion. It explains the behavior of the mobility of the immersed particle in terms of its mass and the viscosity and temperature of the fluid.

(e) In isotropic three-dimensional motion we have a Langevin equation for each component of the velocity:

$$\frac{dV_i}{dt} = -\gamma V_i + S_i(t), \tag{16}$$
$$i = 1, 2, 3,$$

and the mean values of the Langevin force are

$$\langle S_i(t) \rangle = 0, \qquad \langle S_i(t) S_j(t') \rangle = 2D\, \delta_{ij}\, \delta(t-t'). \tag{17}$$

From the energy equipartition law,we have

$$\frac{1}{2} m \langle V_i^2 \rangle = \frac{1}{2} KT$$

and the mean-square displacement for large times is

$$\langle (\vec{X}(t) - \vec{X}_0)^2 \rangle = \sum_{i=1}^{3} \langle (X_i(t) - X_{0i})^2 \rangle = 6 \frac{KT}{\gamma m} t. \tag{18}$$

1.4. Stochastic concepts which arise from the analysis of Brownian motion

I. *STOCHASTIC PROCESSES*

A. Random variable: Let us consider an experiment, under well defined conditions, to measure a certain quantity η. If such a quantity can assume any one of a number of possible numerical values, we say that the quantity is a random or stochastic variable. It will be impossible to predict in advance its value in any given experiment. The statistical characterization of the random variable is given by the probability density W(x), defined as follows :

$$W(x)\,dx = P(\eta \leq x + dx) - P(\eta \leq x)\,. \tag{19}$$

Equation (19) represents the probability of finding the continuous random variable η in the interval $x \leq \eta \leq x + dx$.

B. Random function of time: Let us consider the trajectory of the particle in the fluid (one dimension). The coordinate of the particle is $X(t)$. In Fig. 1 we represent trajectories $X^1(t), X^2(t), X^3(t)$, ... of equal particles under the very same conditions. The curves $X^n(t)$

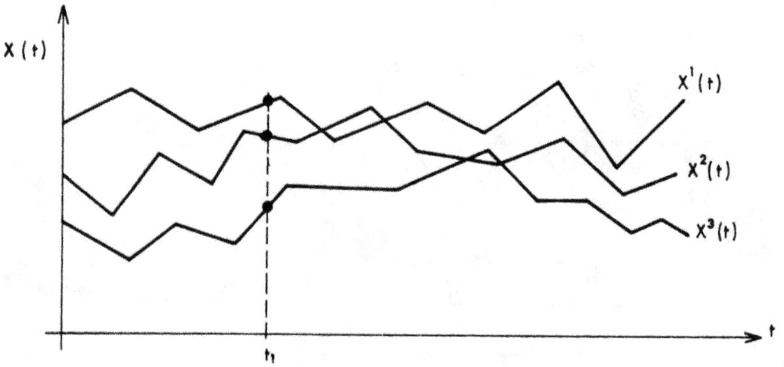

Fig.1. Realizations of a random function of time

are qualitatively similar and unlike each other in detail. Also we can interpret them as corresponding to an ensemble of systems (fluid + particle) and each system leads to a realization $X^n(t)$ of the random function of time $X(t)$. The randomness consists of the fact that the form of each realization $X^n(t)$ changes from one observation to the next in a random manner.

For a fixed time t_1, $\left\{X^n(t_1)\right\}_{n=1,2,\ldots}$ defines a random variable.

C. Stochastic processes: The random variable X(t) defines a stochastic process $\{\eta(t)\}$ which is governed by the Langevin equation

$$\frac{d^2x}{dt^2} = -\gamma\frac{dx}{dt} + S(t)\,.$$

This is a continuous process: $0 < t < \infty$. And the state space is a bounded set of the real line $\mathbb{R}$, where the fluid is located.

D. Mean values: A general fluctuating phenomenon is extremely complex, because it is continually changing in a partly unpredictable manner. The experimenter must perforce renounce total knowledge in favor of some particularly typical values which will give an approximate view: mean values. For instance, $\langle X(t)\rangle$ gives some idea about the order of magnitude of X at the time t. The mean-square value $\langle (X(t) - \langle X(t)\rangle)^2\rangle$ gives information about the dispersion of the values of X(t) about $\langle X(t)\rangle$.

Let us consider the stationary state for Brownian motion. According to Eq. (9), we have for large times

$$V(t) = \int_0^\infty e^{-\gamma\tau} S(t-\tau)\, d\tau\,. \tag{20}$$

From this we can obtain all the moments of the velocity distribution function :

$$\langle V(t)^{2n+1}\rangle = 0\,,$$
$$\langle V(t)^{2n}\rangle = \frac{(2n)!}{2^n n!}\left(\frac{D}{\gamma}\right)^n. \tag{21}$$

To evaluate such moments we need to know the multitime correlation functions of the Langevin force. For this purpose the following assumption is made:

The Langevin force is a Gaussian white noise with zero mean, i.e.,

(a) $\langle S(t)\rangle = 0$; (b) $\langle S(t_1)\, S(t_2)\rangle = 2D\,\delta(t_1 - t_2)$;

(c) $\langle S(t_1)\, S(t_2) \ldots S(t_{2n-1})\rangle = 0$

$$\langle S(t_1)\, S(t_2) \ldots S(t_{2n})\rangle =$$

$$= (2D)^n \sum_{P_d} \delta(t_{i_1} - t_{i_2}) \ldots \delta(t_{i_{2n-1}} - t_{i_{2n}}),$$

where the sum is extended over $(2n)!/(2^n n!)$ permutations .

An important mean value associated to a random function η is the characteristic function:

$$C(u) = \langle e^{iu\eta}\rangle = \int e^{iux}\, W(x)\, dx\,,$$
$$C(u) = 1 + \sum_{n=1}^{\infty} \frac{(iu)^n}{n!} \langle V^n\rangle\,, \tag{22}$$

where W(x) is the distribution function. Again let us consider Brownian motion. The statistical properties of the velocity V(t) in the stationary case are completely determined by the distribution function W(v) because any expectation value can be obtained from W(v) by integration. Since the range of the velocity is $(-\infty, \infty)$, the distribution function is the Fourier transform of the characteristic function:

$$W(v) = \frac{1}{2\pi} \int_{-\infty}^{\infty} C(u)\, e^{-iuv}\, du\,. \tag{23}$$

According to (21) and (22), we have

$$C(u) = \exp\left(-\frac{u^2 D}{2\gamma}\right), \tag{24}$$

$$W(v) = \left(\frac{m}{2\pi KT}\right)^{1/2} \exp\left(-\frac{mv^2}{2KT}\right). \tag{25}$$

The stationary distribution (25) is the Maxwell distribution (Gaussian). Actually, since there is a linear transformation between the velocity and the Langevin force (Eq.(20)) and it is Gaussian, the velocity must be also gaussian. So we can establish that in the Brownian motion the velocity of the particle is a stationary and Gaussian random variable.

II. *STOCHASTIC STATIONARY PROCESSES*

Statistically a time dependent random variable η is characterized by the distribution functions

$$W_1(x_1,t_1)\ ,\ W_2(x_1,t_1;\ x_2,t_2),\ \ldots\ldots, \tag{26}$$

where $W_1(x_1,t_1)dx_1$ is the probability of finding the random variable η in the interval $[x_1,x_1+dx_1]$ for the time $t=t_1$, $W_2(x_1,t_1;x_2,t_2)dx_1dx_2$ is the probability of finding the random variable η in the interval $[x_1,x_1+dx_1]$ for the time $t=t_1$ and in the interval $[x_2,x_2+dx_2]$ for the time $t=t_2$, and so on for the multidimensional distribution function $W_n(x_1,t_1;x_2,t_2;\ldots;x_n.t_n)$.

A stochastic process is stationary if the distribution functions W_n are invariant under time translations: $t_i \rightarrow t_i + T$. The properties of such a process are :

(a) The one-dimensional distribution function is time independent,

$$W_1(x_1,t_1) = W_1(x_1). \tag{27}$$

In fact, $W_1(x_1,t_1) = W_1(x_1,t_1+T)\ \forall\ t_1,T$. By choosing $T = -t_1$ we obtain

$$W_1(x_1,t_1) = W_1(x_1,0) = W_1(x_1).$$

Thus for a stationary process, the mean value of the random variable is constant:

$$\langle \eta(t) \rangle = \int x_1 W_1(x_1,t)\, dx_1 = \int x_1 W_1(x_1)\, dx_1 = ct.$$

In the case of Brownian motion, we have $W_1(x_1) \equiv W(V)$ given by the expression (25) and the mean value of the velocity is constant, $\langle V(t) \rangle = 0$.

(b) The second distribution function depends upon the difference $t_2 - t_1$,

$$W_2(x_1,t_1;x_2,t_2) = W_2(x_1,t_1+T;\ x_2,\ t_2+T).$$

By choosing $T = -t_1$, we obtain

$$W_2(x_1,t_1;x_2,t_2) = W_2(x_1,\ x_2,\ t_2-t_1). \tag{28}$$

So the correlation function of the random variable is also a function of $t_2 - t_1$:

$$\langle \eta(t_1)\, \eta(t_2) \rangle = \iint x_1\, x_2\, W_2(x_1,t_1;\ x_2,\ t_2)\, dx_1\, dx_2$$
$$= F(t_2 - t_1).$$

For Brownian motion we have two stationary random variables, the Langevin force and the velocity:

$$\langle S(t_1)\, S(t_2) \rangle = 2D\, \delta(t_2 - t_1),$$
$$\langle V(t_1)\, V(t_2) \rangle = \frac{D}{\gamma}\, e^{-|t_1 - t_2|}.$$

(c) Analogously, we obtain for the multidimensional distribution function

$$W_n(x_1,t_1;x_2,t_2;\dots;x_n,t_n) = W_n(x_1;x_2,t_2-t_1;\dots;x_n,t_n-t_1).$$

(d) In Brownian motion the velocity of the particle is a stationary random function of time because the causes which generate the process do not change in time. Related to this fact we have the main property of stationary processes: the ergodic property:

{the statistical characteristics of the random process obtained by averaging on its realizations at a given time} = {the characteristics obtained by averaging a single realization for a sufficiently long interval of time}.

So in a stationary stochastic process the different realizations have the very same properties.

The above equality can also be represented as follows:

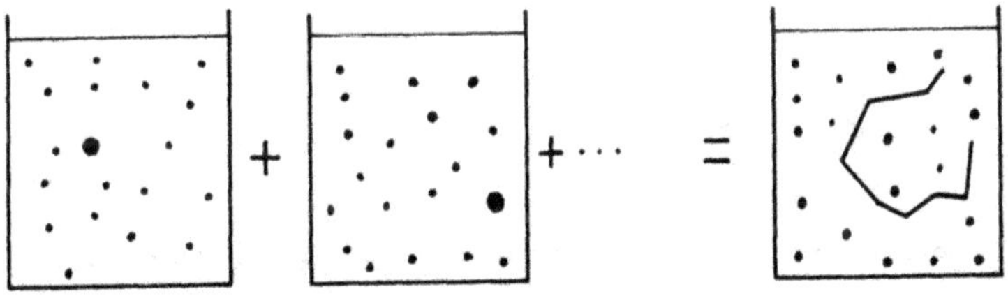

Large numbers of experimenters simultaneously and independently carrying out experiments which are macroscopically identical.

A realization of the motion of the particle in the fluid.

$$\left\{\begin{array}{l}\text{average values}\\ \text{on the ensemble}\end{array}\right\} = \left\{\begin{array}{l}\text{temporal}\\ \text{averages}\end{array}\right\},$$

$$V_0 = \langle V(t) \rangle , \qquad \tilde{V}(T) = \frac{1}{T}\int_t^{t+T} V(t')\,dt' ,$$

$$\tilde{V}(T) \xrightarrow[T\to\infty]{} V_0 .$$

III. <u>*GAUSSIAN PROCESSES*</u>

<u>A</u>. A Gaussian random variable $\{\eta\}$ is characterized by the probability density

$$W(x) = \frac{1}{\sigma\sqrt{2\pi}} \exp\left(-\frac{(x-a)^2}{2\sigma^2}\right), \tag{29}$$

where

$$a = \langle x \rangle = \int_{-\infty}^{\infty} x \, W(x)\,dx , \tag{31}$$

$$\sigma^2 = \langle (x - \langle x \rangle)^2 \rangle . \tag{32}$$

We can also characterize the random variable by its statistical moments

$$b_n = \langle x^n \rangle = \int_{-\infty}^{\infty} x^n \, W(x)\,dx ,$$
$$B_n = \langle (x - \langle x \rangle)^n \rangle = \int_{-\infty}^{\infty} (x - b_1)^n \, W(x)\,dx , \tag{33}$$

where

$$B_2 = \sigma^2,$$
$$B_{2n-1} = 0, \qquad (34)$$
$$B_{2n} = (2n-1)!!\,\sigma^{2n}.$$

So the statistical properties of a Gaussian random variable are completely determined by the mean value $\langle x \rangle$ and the variance σ .

B. A Gaussian random process defined by the time-dependent random variable $\eta(t)$ is such that its statistical properties are completely determined by the mean value $\langle \eta(t) \rangle$ and the correlation function $\langle \eta(t_1)\,\eta(t_2) \rangle$. The odd moments vanish, for example, $\langle \eta(t_1)\eta(t_2)\eta(t_3)\rangle = 0$, and the even moments are constructed in terms of the above correlation function, for example,

$$\langle \eta(t_1)\eta(t_2)\eta(t_3)\eta(t_4)\rangle = \langle \eta(t_1)\eta(t_2)\rangle\langle \eta(t_3)\eta(t_4)\rangle + \langle \eta(t_1)\eta(t_3)\rangle\langle \eta(t_2)\eta(t_4)\rangle + \langle \eta(t_1)\eta(t_4)\rangle\langle \eta(t_2)\eta(t_3)\rangle.$$

IV. *SPECTRAL DENSITY*

The Fourier transform of a random variable $\eta(t)$ is also a random variable:

$$\eta(t) \longrightarrow \tilde{\eta}(\omega) = \int_{-\infty}^{\infty} e^{-i\omega t}\,\eta(t)\,dt. \qquad (35)$$

If the process is stationary we obtain

$$\langle \tilde{\eta}(\omega)\,\tilde{\eta}^*(\omega')\rangle = 2\pi\,\delta(\omega-\omega')\,\mathcal{S}(\omega), \qquad (36)$$

where

$$S(\omega) = 2\int_{-\infty}^{\infty} e^{-i\omega\tau} \langle \eta(\tau)\, \eta^*(0) \rangle \, d\tau \tag{37}$$

is called the spectral density.[4)]

The spectral density of the Langevin force is

$$S(\omega) = 2\int_{-\infty}^{\infty} e^{-i\omega\tau} \langle S(\tau)\, S(0) \rangle \, d\tau = 4D\int_{-\infty}^{\infty} e^{-i\omega\tau} \delta(\tau)\, d\tau = 4D .$$

This is independent of the frequency ω. Therefore the Langevin force associated to Brownian motion is called white noise force.

If the spectral density is ω-dependent, then we have colored noise.

1.5 A mathematical touch

A. Differentiation of random functions

Let us consider a differentiable real function of one variable f(x). Its derivative f'(x) is

$$f'(x) = \lim_{\Delta x \to 0} \frac{f(x+\Delta x) - f(x)}{\Delta x} .$$

The limit requires a "deterministic" concept of the convergence which is not suitable for a random function. In this case the average must included. The generalization to the derivative of a random function $\eta(t)$ is carried out in the following way.[2,5)]

Definition: $\dot{\eta}(t)$ is the derivative of the random function $\eta(t)$ at the instant t if $\eta(t+\Delta t) - \eta(t)/\Delta t \xrightarrow[\Delta t \to 0]{} \dot{\eta}(t)$ in the mean square, i.e.,

$$\lim_{\Delta t \to 0} \langle \left(\frac{\eta(t+\Delta t)-\eta(t)}{\Delta t} - \dot{\eta}(t) \right)^2 \rangle = 0 .$$

In some cases we can only consider the limit for $\Delta t \to 0^+(0^-)$, which means that Δt tends to 0 through positive (negative) values. Then we have the random functions $\dot{\eta}_+(t)$ and $\dot{\eta}_-(t)$, which are called the forward and the backward derivatives.

Two simple properties arise from the above definition:

(a) The averaging and differentiation commute:

$$\frac{d}{dt} \langle \eta(t) \rangle = \langle \frac{d\eta(t)}{dt} \rangle .$$

This property was used in Section 2.

(b) If $\eta(t)$ defines a stationary process, then we have

$$\langle \eta(t_1)\, \eta(t_2) \rangle = F(\tau) \implies \langle \dot{\eta}(t_1)\, \dot{\eta}(t_2) \rangle = -\frac{d^2F(\tau)}{d\tau^2},$$
$$\tau = t_1 - t_2 .$$

B. Integration of random functions

Let us consider a random function $\eta(t)$ defined in the interval $[a,b]$. The integral $I = \int_a^b f(\eta(t),t)\, d\eta(t)$ can be defined in two ways:

Itô Definition: $$I_I = \lim_{\Delta \to 0} \sum_{i=0}^{n-1} f(\eta(t_i), t_i)\, (\eta(t_{i+1}) - \eta(t_i)),$$

where $\Delta = \max (t_{i+1} - t_i)$; $a = t_0 < t_1 < \ldots < t_n = b$. This definition supposes new rules for integration: the Itô calculus.

Stratonovich Definition:

$$I_S = \lim_{\Delta \to 0} \sum_{i=0}^{n-1} f\left(\frac{\eta(t_i)+\eta(t_{i+1})}{2}, \frac{t_i + t_{i+1}}{2} \right) (\eta(t_{i+1}) - \eta(t_i)).$$

It satisfies all the formal rules of an ordinary integral: integration by parts, change of variable and chain rule.

In both definitions the limit must be understood in the mean square. If the function f does not depend on η, then I is an ordinary Stieltjes integral (f(t) continuous) and both definitions agree.

The relevance of Itô and Stratonovich definitions of the stochastic integral is discussed by Van Kampen[6,7] and Smythe et al.[8]

1.6 Stochastic processes related to Brownian motion

A. *ORNSTEIN-UHLENBECK PROCESS*

It is the process which describes the random behavior of the velocity of a small particle in a fluid (Brownian motion): $\{V(t)\}$ by

$$\frac{dV}{dt} = -\gamma V + S(t)$$

or

$$dV = -\gamma V dt + S(t)\, dt .$$

This stochastic process appears in other physical applications.[3]

According to Doob's theorem the only Gaussian stationary Markov process is the Ornstein-Uhlenbeck process.

B. *LANGEVIN PROCESS OR GAUSSIAN WHITE NOISE*

It is a stationary stochastic process having the following properties:

(a) $\langle S(t) \rangle = 0$,

(b) $\langle S(t_1)\, S(t_2) \rangle = 2D\, \delta(t_2 - t_1)$.

(c) The higher moments are given by the rules of the Gaussian process .

The Langevin force defines a Langevin process.

C. *WIENER PROCESS*

It describes the random behavior of the velocity of a particle in a "fluid without viscosity" : $\{V(t)\}$, by

$$\frac{dV}{dt} = s(t),$$
$$dV = s(t)\,dt \equiv dW(t) \equiv dW_t \, , \tag{38}$$
$$V(t) = V(0) + \int_0^t s(t')\,dt' \equiv V(0) + \int_0^t dW(t') \, .$$

The Wiener process is the integral of white noise. We can also say that white noise is the derivative of the Wiener process.

Some properties of the Wiener process are:

(a) $\langle V(t) \rangle = 0$ by choice of the coordinate system .

(b) $\langle V(t)^2 \rangle = 2Dt$ $(\langle S(t')S(t'') \rangle = 2D\,\delta(t'' - t'))$, which expresses that the Wiener process is not stationary.

(c) $\langle (V(t) - V(s))^2 \rangle = 2D(t - s)$, $t > s$,

(d) $V(t_2) - V(t_1)$, $V(t_3) - V(t_2), \ldots, V(t_n) - V(t_{n-1})$ are independent for $t_1 \leq t_2 \leq \ldots \leq t_n$.

(e) Wiener's Theorem:[5] Almost all sample functions of a Wiener process are continuous but nowhere differentiable functions. This means in the case of a particle immersed in a fluid without viscosity that the velocity is a continuous function of time for every realization, but the acceleration of the particle is not defined at any time.

(f) The Wiener process is related to path integration[9] and fractal geometry.[10]

1.7 Markov processes

Markov property: The "future" is independent of the "past" when we know the "present".

This property is the causality principle of classical mechanics carried over to stochastic dynamic systems: the knowledge of the state of a system at a given time s (present) is sufficient to determine the probable state of the system at $t > s$ (future).

Many stochastic processes of theoretical and applied interest possess the Markovian property.

The Ornstein-Uhlenbeck and Wiener processes are Markov processes. This is due to the fact that the differential equations are of first order and the solution is uniquely determined by its initial value, as well as because the Langevin force S(t) is δ-correlated.

The process described by the equation

$$\frac{d\eta}{dt} = f(\eta) + S(t),$$

where $$\langle S(t_1)\, S(t_2) \rangle = \frac{D}{\gamma}\, e^{-\gamma |t_1 - t_2|},$$

is no longer a Markov process. But by introducing a new random variable $\rho(t)$ as follows:

$$\frac{d\eta}{dt} = f(\eta) + \rho, \qquad \frac{d\rho}{dt} = -\gamma\rho + U(t), \qquad \langle U(t_1)\, U(t_2) \rangle = 2D\,\delta(t_1 - t_2),$$

we obtain a bidimensional Markov process.

2. COMPUTER SIMULATIONS OF THE STOCHASTIC EQUATIONS

2.1 Introduction

We can obtain a formal solution of the linear Langevin equation and from it we get the mean values and correlations which characterize Brownian motion. Nevertheless, this approach is not possible in the case of nonlinear Langevin equations:

$$\frac{dV}{dt} = f(V,t) + S(t) \qquad \text{(additive noise)} \tag{39}$$

$$\frac{dV}{dt} = f(V,t) + g(V,t)\,S(t) \qquad \text{(multiplicative noise)} \tag{40}$$

In such cases we can obtain the statistical properties in two ways:

(a) Deriving an equation for the corresponding distribution function: the Fokker-Planck equation or one of its generalizations.

(b) Integrating numerically the stochastic differential equation and computing averages over a set of trajectories.

2.2 The Fokker-Planck equation

The problem of obtaining the statistical properties of the process is reduced to solve a partial differential equation: the Fokker-Planck equation.

The derivation of the Fokker-Planck equation is well presented by Risken[3)] together with methods to solve it.

In these notes, we show as an illustration the Fokker-Planck equation for the Langevin equation:

$$\frac{\partial W(v,t)}{\partial t} = \gamma\,\frac{\partial (vW)}{\partial v} + \frac{\gamma KT}{m}\,\frac{\partial^2 W}{\partial v^2} \tag{41}$$

For a stationary process $\frac{\partial W}{\partial t} = 0$, and since the range of values of the velocity is $(-\infty,\infty)$ we obtain

$$\left(\gamma v + \frac{\gamma KT}{m} \frac{\partial}{\partial v} \right) W = 0 ,$$

from which we get the stationary distribution function for Brownian motion:

$$W(v) = \left(\frac{m}{2\pi KT} \right)^{1/2} \exp\left(- \frac{mv^2}{2KT} \right),$$

which is the Maxwell distribution obtained before, directly from the Langevin equation.

2.3. Computer simulation of the Langevin equation

The methods to integrate stochastic and deterministic differential equations are different. For a deterministic equation the trajectory is uniquely determined by the initial conditions. On the contrary, for a stochastic equation, given any initial condition there is an infinite number of possible trajectories. Every trajectory is related to a realization of the stochastic force:

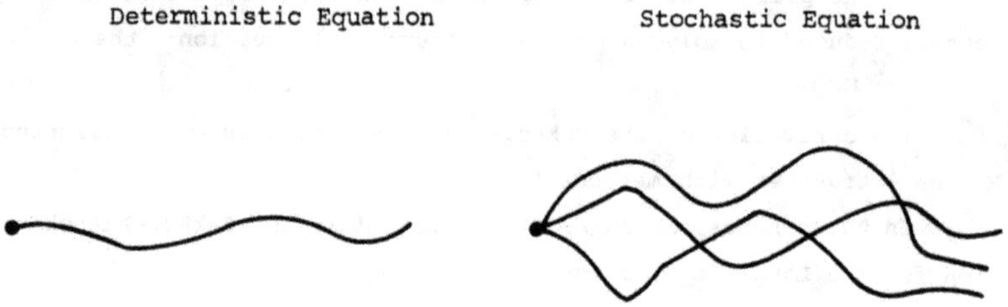

To set up the main features of the numerical integration of a stochastic differential equation, let us consider the nonlinear Langevin equation (39). To integrate numerically the equation in the time interval $[0,T]$, we divide it into N intervals of length $\Delta t = \tau$. And we define

$$V_n = V(n\tau) \quad , \quad S_n = S(n\tau). \tag{42}$$

In the discretization of the Langevin equation we must consider three main steps:

(a) Discretization of the stochastic term S(t)

From the continuous stochastic process S(t) with the statistical properties

$$\langle S(t) \rangle = 0 \quad , \quad \langle S(t_1)\, S(t_2) \rangle = 2D\, \delta(t_2 - t_1),$$

we can generate a discrete stochastic process $\{\varepsilon_n\}$ as follows:

$$\varepsilon_n = \frac{1}{\tau} \int_{n\tau}^{(n+1)\tau} S(t)\, dt \Longrightarrow \begin{array}{l} \langle \varepsilon_n \rangle = 0 , \\ \langle \varepsilon_n \varepsilon_m \rangle = \frac{2D}{\tau} \delta_{nm} , \\ \sigma^2 = 2D/\tau \text{ is the variance.} \end{array} \tag{43}$$

We have that each sequence $\{\varepsilon_n\}$ of random numbers represents a computer simulation of the Langevin force.

The generation of sequences of random numbers and the associated problems belong to an important field of study: Monte Carlo methods.[11)]

In Figs. 2 and 3 we represent a typical numerical simulation of Gaussian white noise.

(b) Interaction between the discretizations of the stochastic and deterministic terms

The numerical simulation of stochastic differential equations and Fokker-Planck equations involve stochastic integrals. Thus, in general, we will obtain different results in the framework of the Stratono-

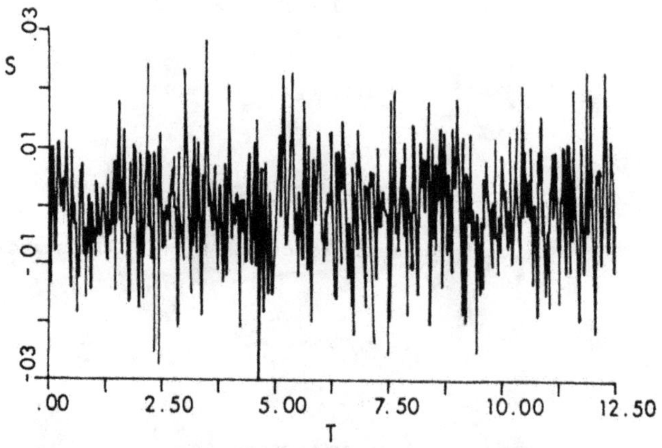

Fig.2. Realization of Gaussian white noise
$\sigma = 0.01$

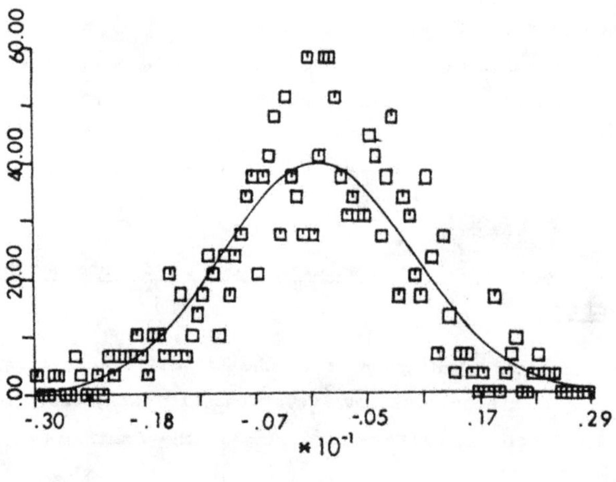

Fig.3. Gaussian distribution of the noise

vich and Itô approaches. This difference appears for the multiplicative noise[3,12] corresponding to Eq. (40). For additive noise (Eq. (39)), the Stratonovich and Itô definitions of stochastic integrals give the same results.

(c) Discretization of the "deterministic term"

A simple discretization is given by Euler's method:

$$\frac{dV}{dt} \longrightarrow V_{n+1} - V_n / \tau , \qquad (45)$$
$$f(V) \longrightarrow f(V_n) .$$

So the simplest numerical scheme for the Langevin equation (39) is

$$V_{n+1} = V_n + \tau f(V_n) + \tau \mathcal{E}_n . \qquad (46)$$

To understand the numerical results obtained from any discretization of a stochastic differential equation, we must take into account the following numerical fact: the numerical schemes associated with a deterministic equation of motion can generate chaos even if the underlying continuous equation does not show chaotic behavior.[13,14] This can falsify the numerical results, due to the superposition of the chaos generated by the numerical scheme onto the externally imposed stochastic noise. For this reason, a basic step is to analyze the features of the scheme associated with the deterministic part of the stochastic differential equation.

Remark. Let us set f = 0 in Eq. (46) and define the random numbers $\eta_n = \sqrt{\tau}\, \mathcal{E}_n$ with

$$\langle \eta_n \rangle = 0 ,$$
$$\langle \eta_n \eta_m \rangle = 2D\, \delta_{nm} . \qquad (47)$$

We obtain

$$V_{n+1} = V_n + \sqrt{\tau}\, \eta_n . \qquad (48)$$

which represents a consistent discretization of the Wiener process (38). The above discretization shows that the samples of the Wiener process are continuous but nowhere differentiable:

$$V_{n+1} \longrightarrow V_n \quad \text{as} \quad \tau \to 0,$$

$$\frac{V_{n+1} - V_n}{\tau} = \frac{\eta_n}{\sqrt{\tau}} \quad \text{it is not defined as} \quad \tau \longrightarrow 0.$$

Thus we have verified numerically Wiener's theorem (!!).

2.4. Chaos generated by the numerical scheme associated with a deterministic differential equation

Let us consider the ordinary differential equation

$$\frac{dx}{dt} = f(x). \tag{49}$$

The simplest finite difference scheme to integrate (49) is the Euler scheme:

$$\frac{x_{n+1} - x_n}{\tau} = f(x_n) \longrightarrow x_{n+1} = x_n + \tau f(x_n). \tag{50}$$

We can interpret the numerical scheme as a nonlinear mapping of the real line $\mathbb{R}$ onto itself:

$$\begin{aligned} T: \mathbb{R} &\longrightarrow \mathbb{R}, \\ x_n &\longrightarrow x_{n+1} = T(x_n). \end{aligned} \tag{51}$$

As a mapping, the numerical scheme can show chaos[13)] even if the underlying continuous equation does not behave chaotically. For this reason, in order to understand such numerical solutions, it is very important to know the properties of the mapping associated to the numerical scheme: fixed points, stability, reversibility, discrete conservation laws,.., etc. As an example of this remark, let us consider the equation [14,15)]

$$\frac{dx}{dt} = x(1-x), \tag{52}$$

whose exact solution is

$$x(t) = \frac{x_0 e^t}{1 - x_0 + x_0 e^t}. \tag{53}$$

The equation has two equilibrium points x = 0 (unstable) and x = 1 (stable).

The central difference scheme associated to equation (52) is

$$\frac{x_{n+1} - x_{n-1}}{2\tau} = x_n (1 - x_n), \tag{54}$$

with initial conditions: X_1 and $X_1 = X_0 + \tau X_0(1 - X_0)$. The numerical solution obtained according to the scheme (54) shows chaotic behavior, and it is independent of the precision in the computations. The long-time behavior of the numerical solution has nothing to do with the solution of the underlying continuous equation. The reason for this behavior lies in the properties of the mapping defined by the central difference scheme[14)]

$$\begin{cases} x_{n+1} = 2\tau x_n (1 - x_n) + y_n \\ y_{n+1} = x_n \end{cases} \longrightarrow \begin{pmatrix} x \\ y \end{pmatrix}_{n+1} = T \begin{pmatrix} x \\ y \end{pmatrix}_n. \tag{55}$$

The fixed points of the mapping T are the equilibrium points of the system (52). But the equilibrium point X = 1 (Y = 1), which is asymptotically stable for the system (52), is an unstable fixed point of the mapping T. So the stability properties of the mapping are not the same as those of the underlying continuous equation. That explains the chaotic behavior of the numerical solution.

The application of the finite difference schemes to approximate the equations of motion of a system does not usually preserve the conservation laws and symmetries of the underlying continuous equations. This means that the solutions may show instabilities, nonphysical properties, etc. For this reason, it is important to construct schemes which respect the dynamics of the underlying continuous equation. As an illustration of this, let us analyze the motion of a classical particle in one dimension under a conservative force:

$$m \frac{d^2x}{dt^2} = F(x) = -\frac{dU}{dx} . \tag{56}$$

The dynamics of the system is contained in the following three features:

(a) Conservation of the energy:

$$E = \frac{1}{2} m \dot{x}^2 + U(x) , \tag{57}$$

$$\frac{dE}{dt} = 0 . \tag{58}$$

(b) Reversibility: invariance under the change $t \rightarrow -t$.

(c) Properties of the potential energy $U(x)$.

<u>*Remark*</u>: If $E = 0 \Longrightarrow \dot{x} = \pm \sqrt{-2\, U(x)/m}$. (59)

Thus the first order system $\dot{x} = f(x)$ can be associated to the motion of a particle in a potential $U(x)$, with total energy $E = 0$.

For a given potential $U(x)$, a numerical scheme which shows[15,16] a discrete version of the properties (a)-(b) is the following:

$$m \frac{X_{n+2} - 2X_{n+1} + X_n}{\tau^2} = -\frac{U(X_{n+2}) - U(X_n)}{X_{n+2} - X_n} . \tag{60}$$

If we multiply (60) by $X_{n+2} - X_n/2\tau$ and we rearrange the terms, we get

$$\frac{E_{n+1} - E_n}{\tau} = 0 \Longrightarrow E_{n+1} = E_n , \; \forall n , \tag{61}$$

$$E_n = \frac{1}{2} m \left(\frac{X_{n+1} - X_n}{\tau}\right)^2 + \frac{1}{2}\left(U(X_{n+1}) + U(X_n)\right), \tag{62}$$

which are consistent discretizations of (57), (58).

According to the above remark, we get the following consistent scheme to solve Eq. (49):

$$X_{n+1} = X_n + \tau \frac{\varepsilon_n}{\sqrt{2}} \left(f^2(X_{n+1}) + f^2(X_n)\right)^{1/2}, \tag{63}$$

where

$$\varepsilon_n = \begin{cases} 1 & \text{if} \quad f(X_n) > 0 \,, \\ 0 & \text{if} \quad f(X_n) = 0 \,, \\ -1 & \text{if} \quad f(X_n) < 0 \,. \end{cases}$$

If the above scheme is applied to Eq. (52), then the numerical solution does not show chaotic behavior.[15]

An extension of the conservative scheme (60) for the two-dimensional motion of a particle is considered by Jiménez and Vázquez[17] in their study of the motion of a charge in a magnetic dipole field.

2.5. Motion of a Charge in a Stochastic Uniform Magnetic Field

As an illustration of a computer simulation of a stochastic differential equation, let us consider the equation which describes the motion of a charge in a stochastic uniform magnetic field[18] (neglecting radiation effects). For this system, some of the relevant expectation values can be obtained from the formal solution as for the Langevin equation. Other mean values can only be obtained numerically. Thus this system offers a good test for any scheme which approximates the corresponding stochastic differential equation.

The equation of motion for a particle of mass m and charge e in a magnetic field $\vec{B}$ is

$$\frac{d\vec{V}}{dt} = \frac{e}{mc}\,\vec{V}\times\vec{B}, \tag{64}$$

where $\vec{V} = d\vec{x}/dt$ and c is the velocity of light.

Let us consider $\vec{B} = (0,0,B)$. We will study the motion of the charge in the plane perpendicular to the magnetic field, so we have $\vec{V}=(V_x,V_y,0)$. By introducing the cyclotron frequency $\omega = eB/mc$, we can write the equation (64) as follows :

$$\frac{dV_x}{dt} = \omega V_y, \qquad \frac{dV_y}{dt} = -\omega V_x. \tag{65}$$

We are going to study the motion of the charge when the magnetic field is a random function of time $B = B_o + B_1(t)$; more precisely, $B_1(t)$ is Gaussian white noise. In terms of ω we have

$$\omega = \omega_o + \varepsilon(t), \quad \langle\varepsilon(t)\rangle = 0, \quad \langle\varepsilon(t)\varepsilon(t')\rangle = 2D\,\delta(t'-t). \tag{66}$$

The equation of motion (65) can be written in complex form as follows:

$$\frac{d\eta}{dt} = -i\,(\omega_o + \varepsilon(t))\,\eta, \tag{67}$$

where $\eta = V_x + iV_y$. The same equation has been used as a model for a random harmonic oscillator.[19)]

The formal solution of (67) reads

$$\eta(t) = \eta(0)\,e^{-i\omega_o t}\,e^{-i\int_0^t \varepsilon(t')\,dt'}, \tag{68}$$

from which we obtain:[3)]

$$
\begin{aligned}
\langle V_x(t)\rangle &= a\, e^{-Dt} \sin(\omega_0 t + \varphi),\\
\langle V_y(t)\rangle &= a\, e^{-Dt} \cos(\omega_0 t + \varphi),\\
\langle V_x^2(t)\rangle &= \frac{E}{2m}\left(1 - e^{-4Dt}\cos(2\omega_0 t + 2\varphi)\right),\\
\langle V_y^2(t)\rangle &= \frac{E}{2m}\left(1 + e^{-4Dt}\cos(2\omega_0 t + 2\varphi)\right),
\end{aligned}
\tag{69}
$$

where the initial conditions are $V_x(0) = a\sin\varphi$, $V_y(0) = a\cos\varphi$, and $E = \frac{1}{2}ma^2$. For other significant mean values, such as $\langle x^2(t)\rangle$, $\langle y^2(t)\rangle$, $\langle R^2(t)\rangle$, and $\langle R(t)\rangle$, it is not possible to obtain closed expressions. For this reason it is worth studying numerically the stochastic equation (64), besides the physical interest in understanding the effect of the simplest stochastic perturbation of the motion of a charge in a uniform and constant magnetic field.

In order to integrate the equation of motion, we use the finite-difference scheme[20)]

$$
\frac{\vec{V}_{n+1} - \vec{V}_n}{\tau} = \frac{e}{mc}\left(\frac{\vec{V}_{n+1} + \vec{V}_n}{2}\right) \times \vec{B}_n , \tag{70}
$$

where $\vec{V}_n = \vec{V}(n\tau)$.

This scheme has a discrete conserved kinetic energy for a general magnetic field, as for the underlying continuous equation:

$$
\frac{1}{2} m \vec{V}_{n+1}^{\,2} = \frac{1}{2} m \vec{V}_n^{\,2} . \tag{71}
$$

A suitable determination of the position coordinates from this scheme is

$$\vec{X}_{n+1} = \vec{X}_n + \frac{1}{2} \left(\vec{V}_{n+1} + \vec{V}_n \right) \tau . \qquad (72)$$

The discretization of the uniform stochastic field is carried out as was indicated in Section 3. In Figs. 4-6 we depict $\langle V_x(t) \rangle$, $\langle V_x^2(t) \rangle$, and $\langle R \rangle$, where $R = (x^2 + y^2)^{1/2}$ is the radius of the trajectory. These figures correspond to the following values of the parameters and initial conditions :

$$\begin{aligned} &V_x(0) = 0 \quad , \quad V_y(0) = 1 \quad , \quad R(0) = 1 , \\ &\omega_0 = 1 \quad , \quad D = 0.005 \quad , \quad \tau = 0.01 . \end{aligned} \qquad (73)$$

What we did in the computation was to simulate 100 particular time evolutions ("trajectories") consistent with the stochastic equation, which correspond to particular realizations of the stochastic magnetic field (a certain sequence of random numbers). And the average of $V_x(t)$, $V_x^2(t)$, and $R(t)$ over their realizations was considered.

The numerical results show a fairly good agreement with the exact solutions (69). On the other hand, $\langle R(t) \rangle$ increases with time, so the stochastic magnetic field destroys the confinement of a charge moving in a uniform constant magnetic field. The effect of the stochastic magnetic field is to spread the trajectory of the particle.

In general the effect of an external noise on a nonlinear system is to change its stability regions. The system exhibits a series of transitions[21,23] which are not expected: *Noise Induced Phase Transitions*.

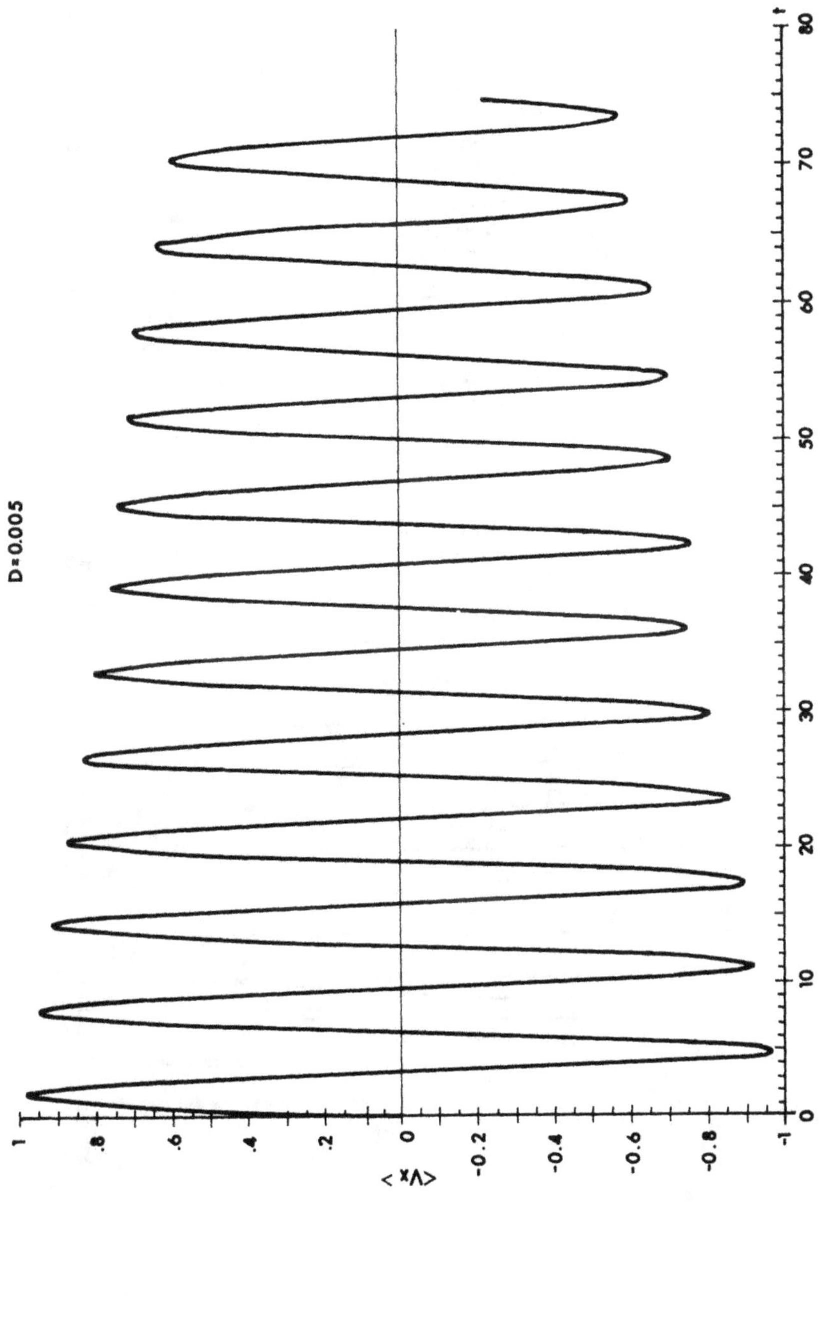

Fig.4. Mean value $\langle V_x(t)\rangle$

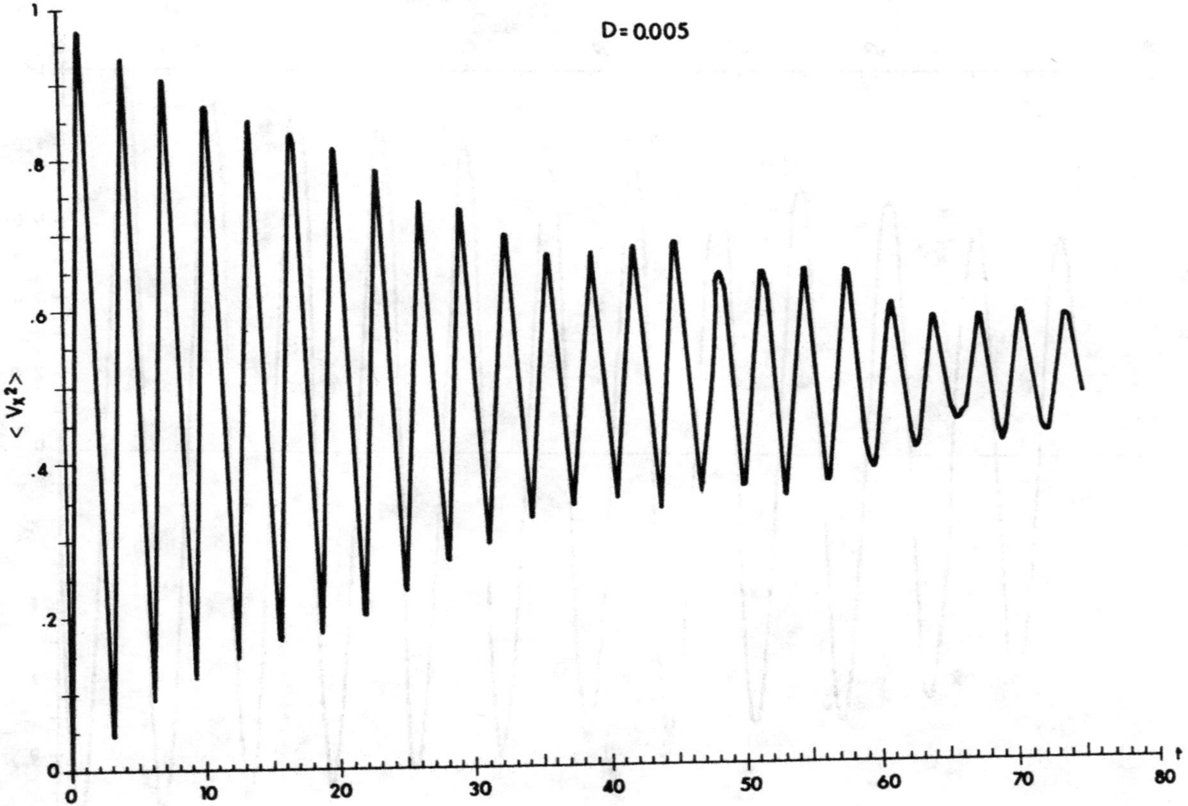

Fig.5. Mean value $\langle V_x^2(t)\rangle$

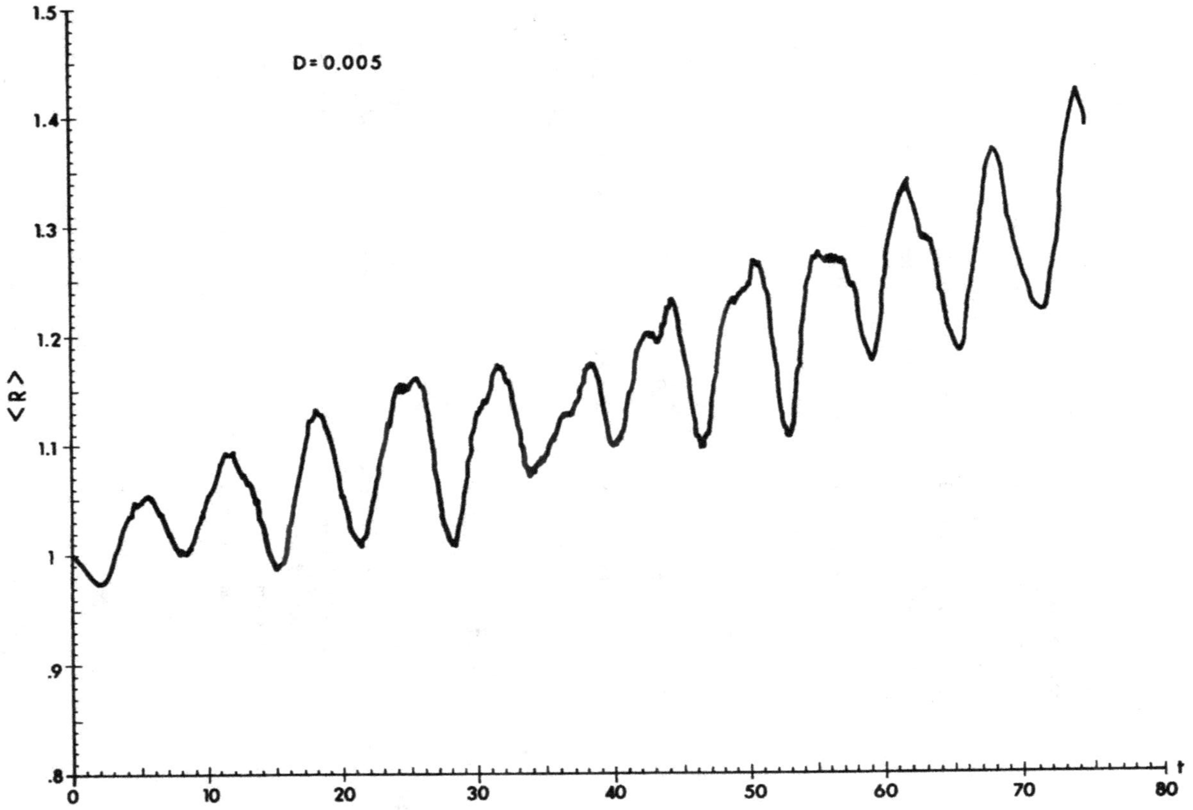

Fig.6. Mean value of the radius $\langle R(t) \rangle$

3. PROPAGATION OF WAVES IN RANDOM MEDIA

3.1. Introduction

The problem of wave propagation in random media has become increasingly important in many areas: acoustics, radio physics, optics, plasma studies, geology, and biology. The applied problems are concerned with the wave propagation (electromagnetic and acoustic waves) and scattering in the atmosphere, the ocean, the earth, biological media, etc. These media are, in general, randomly varying in time and space, so the amplitude and phase of the waves fluctuate randomly in time and space.[24-26]

All problems are characterized by statistical descriptions of the waves and media. Thus in the framework of stochastic processes, it is possible in principle to develop a basic formulation common to all these problems.

3.2. The linear scalar wave equation in one space dimension

As an illustration of the method to study stochastic wave equations, let us consider the simplest one - the linear scalar wave equation in one space dimension:

$$\varphi_{tt} = c^2 \varphi_{xx} . \tag{74}$$

It represents longitudinal sound waves in air, in which the air molecules vibrate about their mean positions in the direction of wave propagation. It also appears in optics. The equation determines the spatial and temporal evolution of the amplitude φ in a homogeneous, isotropic, and conservative system (c = constant).

The energy and momentum transported by the perturbation are following

$$E = \frac{1}{2} \int_{-\infty}^{\infty} (\varphi_t^2 + c^2 \varphi_x^2) \, dx ,$$

$$P = - c^2 \int_{-\infty}^{\infty} \varphi_x \varphi_t \, dx , \qquad (75)$$

$V = \frac{P}{E}$, velocity of the center of energy associated to the perturbation.

For an inhomogeneous medium characterized by a refraction index n(x), the wave equation (74) takes the form

$$\varphi_{xx} - \frac{n^2(x)}{c^2} \varphi_{tt} = 0 . \qquad (76)$$

If the properties of the medium are stochastic, then n(x) is a random function of position. In this case, Eq. (76) is one of the simplest stochastic wave equations. Let us assume that

$$n^2(x) = n_0^2 (1 + \alpha \eta(x)) , \qquad (77)$$

$$\langle \eta(x) \rangle = 0 .$$

Then we obtain for the stationary solutions (deformed plane waves):

$$\varphi(x,t) = e^{-i\omega t} \varphi(x) , \qquad (78)$$

$$\varphi_{xx} + \frac{\omega^2 n_0^2}{c^2} \varphi = - \frac{\omega^2 n_0^2}{c^2} \alpha \eta(x) \varphi . \qquad (79)$$

By absorbing the factor $\omega n_0/c$ in the variable x, so that the distance is measured in unperturbed wave lengths, we get

$$\varphi_{xx} + \varphi = - \alpha \eta(x) \varphi . \qquad (80)$$

The random term enters as a coefficient (multiplicative noise). This is the root of all the mathematical difficulties of the theory, since we do not know how to find an exact solution of such a wave equation. Let us remark that the stochastic wave equation would be more complicated if the refraction index were also random in time.

3.3. Approaches to solving the stochastic wave equation

A. For a static and stochastic refraction index, the stationary solutions satisfy the stochastic ordinary differential equation (80). In this way we can apply the methods to solve the Langevin equations.[19] An approximate equation can be obtained for the average amplitude:

$$\frac{d^2}{dx^2}\langle \varphi(x)\rangle + \alpha^2 C_2 \frac{d}{dx}\langle \varphi(x)\rangle + \left(1-\frac{1}{2}\alpha^2 C_1\right)\langle \varphi(x)\rangle = 0, \tag{81}$$

where

$$C_1 = \int_0^\infty \langle \eta(x)\,\eta(x-z)\rangle \sin 2z \, dz,$$

$$C_2 = \int_0^\infty \langle \eta(x)\,\eta(x-z)\rangle (1-\cos 2z)\, dz.$$

Equation (81) shows that there are fluctuations in the frequency and a damping of the average amplitude. These two effects are analogous to the energy shift and natural broadening in the theory of scattering.

B. Method of small perturbations[19,27,28]

Let us consider the scalar wave equation in three space dimensions. For the stationary solutions we have

$$\Delta \varphi + \varphi = -\alpha\, \eta(\vec{x})\, \varphi. \tag{82}$$

The solution of Eq. (82) can be represented as

$$\varphi(\vec{r}) = \varphi_0(\vec{r}) - \alpha \int G(\vec{r},\vec{r}')\, \eta(\vec{r}')\, \varphi(\vec{r}')\, d^3\vec{r}', \quad (83)$$

where $\varphi_0(\vec{r})$ is the solution of the unperturbed equation and represents the wave that would be propagated in the absence of the stochastic perturbation, and $G(\vec{r},\vec{r}')$ is the Green's function for the homogeneous medium. We can solve (83) by iteration; thus for the first order we obtain

$$\varphi(\vec{r}) = \varphi_0(\vec{r}) - \alpha \int G(\vec{r},\vec{r}')\, \eta(\vec{r}')\, \varphi_0(\vec{r}')\, d^3\vec{r}'. \quad (84)$$

This is the single-scattering approximation used in optical problems by Rayleigh, and which is often called the Born approximation in quantum mechanics. In this approximation the amplitude φ_0 of the incident wave is not changed by the stochastic medium, while the scattered field depends linearly on $\eta(\vec{x})$.

From Eq. (83) we can obtain an equation which is a second order approximation for the average of the amplitude:

$$(\Delta + 1)\langle \varphi(\vec{r})\rangle = \alpha^2 \int \langle \varphi(\vec{r})\, G(\vec{r},\vec{r}')\, \varphi(\vec{r}')\rangle \langle \varphi(\vec{r}')\rangle\, d^3\vec{r}'. \quad (85)$$

C. Path integral methods

The application of the path integral techniques to the study of wave propagation in random media is described in the book by Schulman.[9)]

D. Numerical methods

These methods assume the availability of a powerful computer. In Section 3.5 we show an example of such numerical simulations. The method is the same as the one indicated in Part II, for the numerical simulation of a Langevin equation. Now the difference is that we have a stochastic partial differential equation which can be nonlinear. The method consists in finding a suitable numerical scheme to integrate the equation for different realizations of the stochastic term and to compute averages over a certain number of realizations.

3.4. The stochastic Schrödinger equation

The nonrelativistic motion of a quantum particle in a random potential is described by the stochastic Schrödinger equation

$$i\hbar \frac{\partial \Psi}{\partial t} = -\frac{\hbar^2}{2m} \Delta \Psi + V(\vec{x}) \Psi , \tag{86}$$

where $V(\vec{x})$ is a random function. The equation describes the motion of electrons in disordered systems like crystals with impurities and liquid metals.[29–31]

In Eq.(86) the noise is multiplicative, but unlike wave equation (76) it does not appear in the derivative term. For the stationary solutions we have

$$\Psi = e^{-iEt/\hbar} \psi(\vec{x}), \tag{87}$$

$$\left(-\frac{\hbar^2}{2m} \Delta + V(\vec{x})\right) \psi = E \psi . \tag{88}$$

This is an eigenvalue problem in a random potential. At present, the study of the spectral properties of random Schrödinger equations and other random wave equations[32,33] is a very intensive field of work.

An important phenomenon related to Eq. (88) is the Anderson localization, as it is called in condensed matter physics. In some conditions of disorder and in particular for any disorder in one dimension, all the eigenfunctions of the Schrödinger equation with a random potential are exponentially localized for almost every realization of the random potential.[29,33,34] In such cases the system behaves as an insulator. The opposite behavior occurs when the eigenfunctions are not localized.

3.5. Behavior of a Sine-Gordon soliton under stochastic perturbations

As an illustration of the numerical approach to study wave equations in random media, let us consider the Sine-Gordon equation with stochastic perturbations. In particular, we are going to summarize the details of the method which is described in Refs. [35-37]. The study is carried out for the evolution of a single soliton in a random medium. When the stochastic perturbation is very small, a perturbative approach[36,38] is developped.

The unperturbed Sine-Gordon equation is

$$\varphi_{tt} - \varphi_{xx} + \sin\varphi = 0. \tag{89}$$

The soliton-antisoliton solutions are

$$\varphi_{\pm} = 4\tan^{-1}\{\exp[\pm\gamma(x - X)]\}, \tag{90}$$

where $\gamma = (1 - U^2)^{-1/2}$, U is the velocity, and X locates the center of the soliton, being $X = X_o + Ut$. There are two main interpretations of Sine-Gordon solitons:[39]

(a) Structures with remarkable stability properties which appear in different problems: Josephson transmission lines, spin waves in ^{3}He, etc.

(b) A model of a relativistic extended particle.

The behavior of solitons under stochastic perturbations is of considerable interest for applications. The noise can simulate the interaction with an external fluctuating field, thermal noise, or spatial inhomogeneities.

In order to understand the effect of the noise on a soliton, we must determine the region of the strong and weak noises. A strong noise destroys the solitonic entity, while under a weak noise the soliton is preserved. The fluctuations in the velocity, position, and amplitude are the relevant effects. To find out such regions we must fix the reference times. A characteristic time associated with the soliton was defined as follows:

$\tau_0 = 4$: time for propagation of a light signal through the soliton

and the computation time was

$$T = 3\,\tau_0$$

We studied the perturbed Sine-Gordon equation[36)]

$$\varphi_{tt} - \varphi_{xx} + \sin\varphi + \alpha\,\varphi_t + V(x,t)\,\varphi + F(x,t) = 0, \tag{91}$$

where V and F are functions localized in space and varying randomly in time, while $\alpha\,\varphi_t$ represents a loss term. The term F(x,t) represents an additive noise, while V(x,t) corresponds to a multiplicative noise.

To accomplish the numerical integration of equation (91), the following finite-difference scheme was used:

$$\frac{\varphi_e^{n+1} - 2\varphi_e^{n} + \varphi_e^{n-1}}{(\Delta t)^2} - \frac{\varphi_{e+1}^{n} - 2\varphi_e^{n} + \varphi_{e-1}^{n}}{(\Delta x)^2} - \frac{\cos\varphi_e^{n+1} - \cos\varphi_e^{n-1}}{\varphi_e^{n+1} - \varphi_e^{n-1}} +$$

$$+ \alpha\,\frac{\varphi_e^{n+1} - \varphi_e^{n-1}}{2\Delta t} + \frac{1}{4}\left(V_e^{n+1} + V_e^{n-1}\right)\left(\varphi_e^{n+1} + \varphi_e^{n-1}\right) +$$

$$+ \frac{1}{2}\left(F_e^{n+1} + F_e^{n-1}\right) = 0. \tag{92}$$

This scheme is explicit and at each time step a simple functional equation needs to be solved for the unknown φ_ℓ^{n+1}. For the unperturbed Sine-Gordon equation there is a discrete energy which is constant:

$$E^n = \frac{1}{2}\sum_\ell \Delta x \left(\frac{\varphi_\ell^{n+1}-\varphi_\ell^n}{\Delta t}\right)^2 + \frac{1}{2}\sum_\ell \Delta x\left(\frac{\varphi_{\ell+1}^{n+1}-\varphi_\ell^{n+1}}{\Delta x}\right)\left(\frac{\varphi_{\ell+1}^{n}-\varphi_\ell^{n}}{\Delta x}\right)$$

$$+ \sum \Delta x \left(1 - \frac{\cos\varphi_\ell^{n+1}+\cos\varphi_\ell^n}{2}\right). \tag{93}$$

In this case the stability and convergence of the scheme has been studied by Guo Ben Yu et al.[40] Related to this is the general open problem of proving the stability and convergence of numerical schemes associated to stochastic partial differential equations.[41]

For the case of weak perturbations, a perturbative approach is possible by assuming that their predominant effect is to modulate the center and the velocity of the soliton. Thus we simplify the problem of solving a set of ordinary stochastic differential equations.

Numerical results are described in Refs. [35] and [36].

ACKNOWLEDGMENTS

The preparation of this course benefited from the stimulating research atmosphere of the BiBos Project at Bielefeld University.

REFERENCES

1. Nelson, E., Dynamical Theories of Brownian Motion (Princeton University Press, Princeton, 1967).
2. Arnold, L., Stochastic Differential Equations: Theory and Applications (Wiley, New York, 1974).
3. Risken, H., The Fokker-Planck Equation: Methods of Solution and Applications (Springer, Berlin, 1984).
4. Rice, S.O., "Mathematical Analysis of Random Noise," in Selected Papers on Noise and Stochastic Processes, edited by N. Wax (Dover, New York, 1954).
5. Hoel, P.G., Port, S.C., and Stone, C.J., Introduction to Stochastic Processes (Houghton-Mifflin, Boston, 1972).
6. Van Kampen, N.G., "Itô Versus Stratonovich," J. Stat. Phys. 24, 175-187 (1981).
7. Van Kampen, N.G., "The Validity of Nonlinear Langevin Equations," J. Stat. Phys. 25, 431-442 (1981).
8. Smythe, J., Mors, F., McClintock, P.V.E., and Clarkson, D., "Itô Versus Stratonovich Revised," Phys. Lett. 97A, 95-98 (1983).
9. Schulman, L.S., Techniques and Applications of Path Integration (Wiley, New York, 1981).
10. Mandelbrot, B.B., The Fractal Geometry of Nature (Freeman, San Francisco, 1983).
11. Hammersley, J.M., and Handscomb, D.C., Monte Carlo Methods (Chapman and Hall, London, 1979).
12. Sancho, J.M., San Miguel, M., Katz, S.L., and Gunton, J.D., "Analytical and Numerical Studies of Multiplicative Noise," Phys. Rev. A26, 1589-1609 (1982).
13. Gumonwski, I., and Mira, C., Dynamique Chaotique: Transformations Ponctuelles. Transition Ordre-Désordre (Cepadues Editions, 1980).
14. Yamaguti, H., and Ushiki, S., "Chaos in Numerical Analysis of Ordinary Differential Equations," Physica 3D, 618-626 (1981).
15. Vázquez, L., "Long Time Behavior in Numerical Solutions of Certain Dynamical Systems," An. Fis. (to be published).
16. Pascual, P.J., and Vázquez, L., "A Numerical Scheme for One Dimensional Mechanics Problems," Had. J. (to be published).

17. Jiménez, S., and Vázquez, L., "Motion of a Charge in a Magnetic Dipole Field I: Painlevé Analysis and a Conservative Numerical Scheme," BiBos 225/96 (BiBos: Bielefeld-Bochum-Stochastik Project).

18. Pascual, P.J., and Vázquez, L., "Motion of a Charge in a Stochastic Uniform Magnetic Field," BiBos 224/86.

19. Van Kampen, N.G., "Stochastic Differential Equations," Phys. Rep. 24, 171-228 (1976), and references therein.

20. Potter, D., Computational Physics (Wiley-Interscience, Chicester, 1977).

21. Horsthemke, W. and Lefever, R., "Phase Transition Induced by External Noise," Phys. Lett. 64A, 19-21 (1977).

22. Stochastic Nonlinear Systems in Physics, Chemistry and Biology, edited by L. Arnold and R. Lefever (Springer, Berlin, 1981).

23. Horsthemke, W., and Lefever, R., Noise-Induced Transition Theory and Applications in Physics, Chemistry and Biology (Springer, Berlin, 1983).

24. Ishimaru, A., Wave Propagation and Scattering in Random Media, Vols. I, II (Academic, New York, 1978).

25. Chernov, L.A., Wave Propagation in a Random Media (McGraw-Hill, New York, 1960).

26. Herraiz, M. and Espinosa, A.F., "Scattering and Attenuation of High-Frequency Seismic Waves: Development of the Theory of Coda Waves," Open-file Report 86-455 (U.S.Geol. Surv., Denver, Colorado), and references therein.

27. Barabanenkov, Yu. N., Kravtsov, Yu. A., Rytov, S.M., and Tamarskii, V.I., "Status of the Theory of Propagation of Waves in a Randomly Inhomogeneous Medium," Soviet Physics Uspecki 13, 551-680 (1971).

28. Jeffrey, A. and Kawahara, T., Asymptotic Methods in Nonlinear Wave Theory (Pitman, Boston, 1982).

29. Ishii, K., "Localization of Eigenstates and Transport Phenomena in the One-Dimensional Disordered Systems," Suppl. Prog. Theor. Phys. 53, 77-138 (1973).

30. Ouchinnikov, A.A. and Erikhman, N.S., "Motion of a Quantum Particle in a Stochastic Medium," Sov. Phys. JETP 40, 733-737 (1975).

31. Thouless, D.J., "Electrons in Disordered Systems and the Theory of Localization," Phys. Rep. 13, 93-142 (1974).

32. Souillard, B, "Electrons in Random and Almost Periodic Potentials", Phys. Rep. 103, 41-46 (1984): "Spectral Properties of Discrete and Random Schrodinger Operators: A Review," preprint for the Proceedings of the Meeting on Random Media of the IMA, Minnesota (1985).

33. Martinelli, F., "A Rigorous Analysis of Anderson Localization," BiBos 152/85.

34. Souillard, B., "Fractals and Localization," in Amorphous and Liquid Materials, Proceedings of the NATO ASI, edited by E. Lutscher, G. Fritsch, and G. Jacucci (Nijhoff, Dordrecht, Netherlands, 1987, pp. 19-28.

35. Marchesoni, F., and Vázquez, L., "Sine-Gordon Solitons in the Presence of a Noisy Potential," Physica 14D, 273-276 (1985).

36. Pascual, P.J., and Vázquez, L., "Sine-Gordon Solitons under Weak Stochastic Perturbations," Phys. Rev. B32, 8305-8311 (1985).

37. Rodríguez, M.J., and Vázquez, L., "Behavior of the φ^4-Kinks Under Stochastic Perturbations," in preparation.

38. Bass, F.G., Konotop, V.V. and Sinitsyn, Yu. A., "Solitons in a Random Force Field," Sov. Phys. JETP 61, 318-322 (1985).

39. Dodd, R.K., Eilbeck, J.C., Gibbon, J.D. and Morris, H.C., Solitons and Nonlinear Wave Equations (Academic, London, 1982).

40. Guo Ben-Yu, Pascual, P.J., Rodríguez, M.J., and Vázquez, L., "Numerical Solution of the Sine-Gordon Equation," Appl. Math. Comp. 18, 1-14 (1986).

41. Helfand, E., "Numerical Integration of Stochastic Differential Equations," Bell Sys. Tech. J. 58, 2289-2299 (1979).

CHAOS IN SOLID STATE SYSTEMS

A. Zettl
Department of Physics
University of California, Berkeley
Berkeley, Calif. 94720 U.S.A.

An introduction to chaotic dynamics in driven solid state systems is presented. First, methods of chaos identification and analysis are described. The methods are then applied to driven Josephson junctions, charge density waves, and semiconductors. Both temporal and spatial chaos are investigated. The sensitivity of the dynamical response to external noise, far from or near to a dynamical instability, is also discussed.

CONTENTS

1. INTRODUCTION

The inherent complexity of the "turbulent" dynamical response of fluids, flame fronts, or even biological systems has long been appreciated. Recent advances in the theory of chaos have generated a new understanding of and appreciation for complex nonlinear dynamics. Methods of analysis have been developed which allow direct determination of important system parameters (such as number of active degrees of freedom) from the seemingly erratic response. The richness of classical dynamics in the chaotic regime has suggested interesting parallels with well established theories of quantum dynamics.

Chaotic dynamics appears to be a very general phenomenon. In this brief review, we shall explore chaotic dynamics in solid state systems. "Solid state turbulence" is a relatively new field of study and only a handful of chaotic solid state systems have actually been identified. Nevertheless, a wealth of experimental data and theoretical results exist. This paper is not meant to be an exhaustive review of that work, but merely to serve as an introduction to this exciting and rapidly growing field.

Three classes of solid state systems form the focus of this paper: Josephson junctions, charge density waves, and semiconductors (including p-n junctions, photoconductors, and electron-hole plasmas). In general the equations of motion for these systems are derived from a quantum framework, but have simple classical analogs. An attempt has been made to keep the "chaos" language as simple as possible and to focus only on the features of the solid state systems directly relevant to the chaotic response.

2. CHAOTIC RESPONSE-- IDENTIFICATION AND ANALYSIS

2.1 A Hypothetical Experiment

Let us consider a "black box" system as shown in Fig. 1. The system may be a complicated one, with many degrees of freedom. The

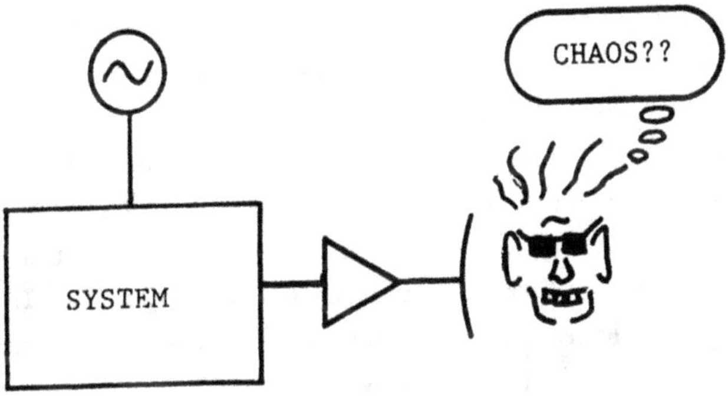

Figure 1. Testing a system for chaos.

exact equations of motion are unknown, but it is assumed that the system is dissipative. The box has one input port and one exit port. The input port is connected to a sine wave oscillator; the exit port is connected to an audio amplifier and Hi-Fi speaker. We listen to the speaker. With the oscillator amplitude knob set to zero, we hear nothing. As the amplitude is steadily increased, we hear first a relatively pure tone corresponding to the oscillator frequency at ω_{ex}, then also higher harmonics, then also a lower frequency, and then loud, scratchy noise. Have we discovered chaos in the box? How many active degrees of freedom does the system possess? Can we determine this and more from just the single output variable, the acoustic wave amplitude versus time? Would it be advantageous to have more than one output port on the box, each corresponding to a different "part" of the system? In this and the following sections, we shall develop concepts and methods which address these questions, and apply them to selected solid state systems.

2.2 Chaotic Dynamics

We here treat a dynamical system with dissipation and potentially many degrees of freedom. The system is self-driven, externally driven, or both. The temporal evolution of the system is represented as a trajectory of a point in multidimensional phase space. Figure 2 shows one such possible trajectory in a three dimensional phase space.

Different initial conditions will correspond to different deterministic trajectories in the phase space. Since the system is dissipative, the volume occupied in phase space by the trajectories will converge to a limit set, i.e. to an <u>attractor</u>.[1] If the system is <u>chaotic</u>, the attractor to which the trajectories converge is a <u>strange attractor</u>.[2] Figure 3 shows a typical strange attractor. The

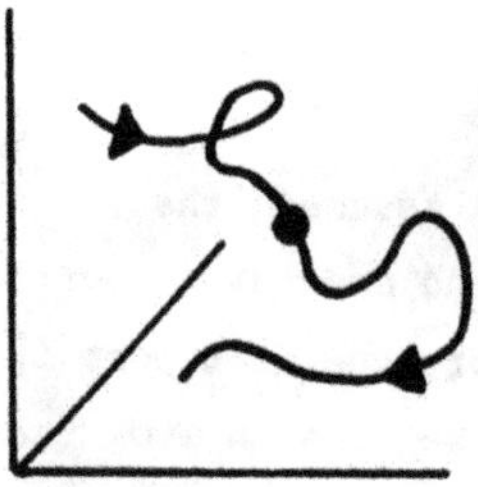

Figure 2. Trajectory of system in phase space.

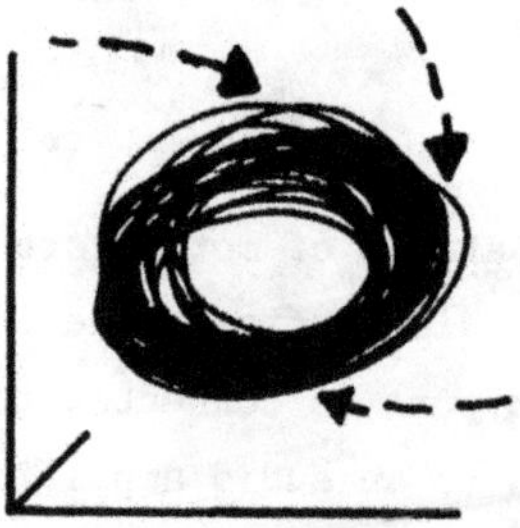

Figure 3. Attractor in phase space.

strange attractor has some interesting properties, which result from the multiple foldings in phase space which the attractor has undergone. If particular, on a strange attractor nearby trajectories simultaneously diverge exponentially from one another, but remain bounded in phase space. the rate of divergence (or convergence) of the trajectories is measured by the Lyapunov exponents.[3] For chaos to exist, there must be at least one positive Lyapunov exponent (i.e., there must be divergence present).

The fact that nearby trajectories on a strange attractor may diverge leads to a sensitive dependence on initial conditions. This means that if, at a given time, all the phase space coordinates are known exactly (possible only in a classical framework), then the evolution of the system is fully determined. However, an infinitesimal uncertainty in even one of the coordinates leads to an exponential growth in the uncertainty of the trajectory. This implies that even if the system returns almost exactly to the same initial condition (near recurrence), the resulting orbit in phase space will not be almost periodic. Rather, the response will sample nearly all orbits, leading to broadband (noisy) response. This gives us our first test for the existence of chaos: a broadband response spectrum.

In the hypothetical experiment of Section 2.1, we could have connected the output of the amplifier to a spectrum analyzer (Fourier transform); the resulting power spectra might be as shown in Fig. 4. In Fig. 4d, broadband noise is observed. Does this mean that the system is chaotic? The answer is probably yes, but more analysis is necessary. Broadband response is a signature, but not proof, of chaos.

Another important feature of a strange attractor is that it has a fractal dimension.[3] In a sense, the attractor has a certain nonintegral "roughness" to it, which can always be recovered by a suitable change of magnification scale. The utility of determining the dimension of a strange attractor (other than insuring that the attractor is indeed strange) is that the dimension gives approximately the number of active degrees of freedom in the system. For example, a

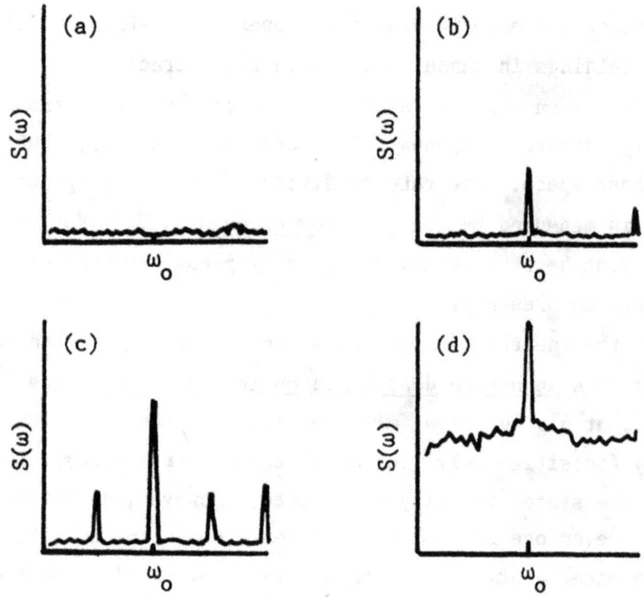

Figure 4. Response power spectra for four different drive amplitudes. a) zero drive; b) drive = 1; c) drive = 2, period doubling; d) drive = 3, chaos.

turbulent fluid might be thought to have essentially an infinite number of degrees of freedom, but the deterministic chaotic response might be described by only a few active degrees of freedom. This is the meaning of "low dimensional chaos".

How does one experimentally determine the dimension of the attractor? The attractor is in general a complicated object residing in multidimensional phase space. A simplification occurs in viewing the attractor if it is projected onto a two dimensional space, defined by two independent variables, thus creating a <u>phase portrait</u>.[1-4] The independent variables might be, for example, the voltage V(t) supplied by the drive oscillator and the current response I(t) of the system. Two possible phase portraits are shown in Fig. 5. In Fig. 5b, the orbit is periodic. The phase portrait gives a projected view of the attractor. Dimensional information can be obtained from a <u>Poincaré section</u>,[1-4] generated by placing a two-dimensional surface in multi-dimensional phase space and recording on the surface (with a dot, say) each time the trajectory unidirectionally pierces the surface. In practice this can be accomplished by "strobing" the phase portrait, using as the strobe trigger the drive oscillator frequency or a dominant frequency internal to the system. Figure 6 shows Poincaré sections corresponding to the phase portrait of Fig. 5.

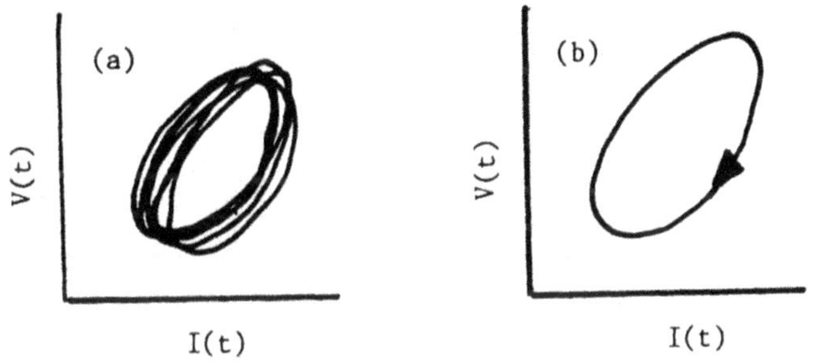

Figure 5. Current response versus voltage drive, yielding phase portrait.

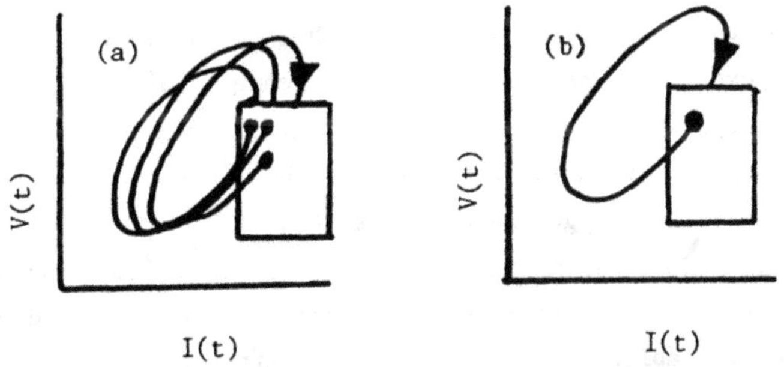

Figure 6. Poincaré section of attractor.

The Poincaré section gives a slice, rather than a projection of the attractor. The <u>return map</u>[1,4] relates the state of the system at index n+1 (n could be a time unit) to the state of the system at index n. The return map of the system described in Fig. 6b, for example, is expecially simple. As shown in Fig. 7a, the return map here consists of only one point. Figure 7b shows a more complex return map. Note that the points on this

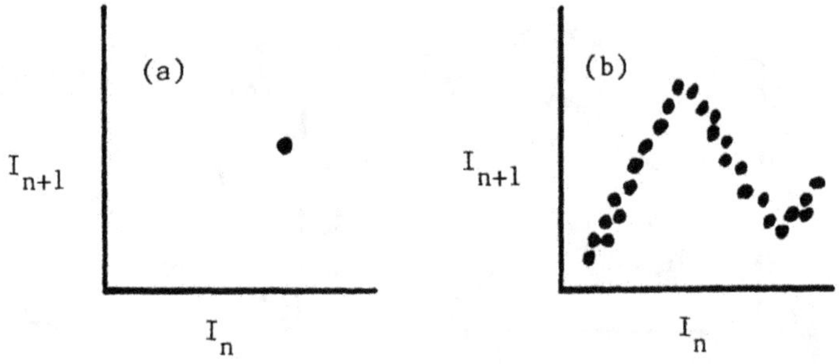

Figure 7. Return map.

mapping fall more or less on a continuous line; this is an example of a one-dimensional return map. The return map in general corresponds to successive iterates of the system. If the return map for a system is known, its evolution can be tracked "geometrically" using only a straight edge. This is illustrated in Fig. 8. Iterating a map is in general far easier than integrating the (often highly complex) differential equations governing the system.

How does one experimentally generate a Poincaré section or return map? To extract the Poincaré section, one needs only to display two independent variables of the system (say voltage and current) on the vertical and horizontal axes of an oscilloscope, and strobe the oscilloscope with an appropriate frequency trigger. The return map

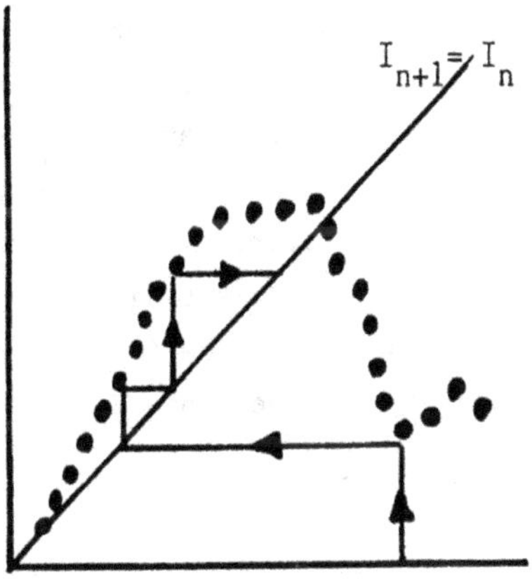

Figure 8. Iterating a map.

can, rather surprisingly, be generated or reconstructed from only a single variable of the system.[5,6] If the output of the system is converted to a voltage V(t), then the time series $V(t_i + T)$, $V(t_i + 2T)$, $V(t_i + 3T)$,..., $V(t_i + (m-1)T)$ defines a return map, provided $m > D$, with D the embedding dimension. This type of reconstruction of phase space is a particularly powerful experimental tool.

The fractal dimension d_F of the attractor can be extracted from the Poincaré section or from the reconstructed phase space (return map). A number of box-counting algorithms have been suggested. Strictly speaking, the fractal dimension can be expressed as[3)]

$$d_F = \lim_{\varepsilon \to 0} \frac{\log M(\varepsilon)}{\log (1/\varepsilon)} , \qquad (2.1)$$

where the phase space has been divided up into P-dimensional cubes of side ε, where P is the dimension of the phase space in which the attractor lies. $M(\varepsilon)$ is the number of cubes visited by the attractor.

Unfortunately, Eq. (2.1) is difficult to apply to an experimental situation. A useful practical technique is to reconstruct the P-dimensional phase space from a single variable, and then calculate the pointwise dimension, or natural measure.[3] One then assumes that the pointwise dimension is a good approximation to d_F. The pointwise dimension is determined by randomly selecting a point on the attractor, and then placing a P-dimensional hypercube (or hypersphere) of radius r around that point. The number of other points N(r) from the attractor within that cube is then evaluated. With

$$N(r) \sim r^{d_F} , \tag{2.2}$$

a plot of log N(r) versus log r yields a straight line of slope d_F. An important distinction between the pointwise dimension and d_F is that d_F is a global dimension (refers to the entire attractor), whereas the pointwise dimension is a local quantity. Fortunately, many attractors have a uniform dimension throughout, and thus the the local measure is equivalent to the global measure.

The topological nature of the attractor is dictated by the parameters defining the system. Often, changes in these parameters drastically alter the system and hence also the attractor. We may define a parameter λ as a control parameter; of interest is how the system and its attractor are changed as λ is changed. For example, in the sequence of Fig. 4, λ might be the oscillator amplitude voltage. If we were to plot the output of the system (say the current response) as a function of time, then the traces of Fig. 9 might result. Note that local minima in the current response have in Fig. 9 been underlined. These minima define a bifurcation diagram.[1-4] The bifurcation diagram expresses, for example, the local minima versus the control parameter. An example is given in Fig. 10, where λ_0, λ_1, λ_2, and λ_3 are identified. At each bifurcation (splitting)point, the system period doubles, and the topological nature of the attractor is

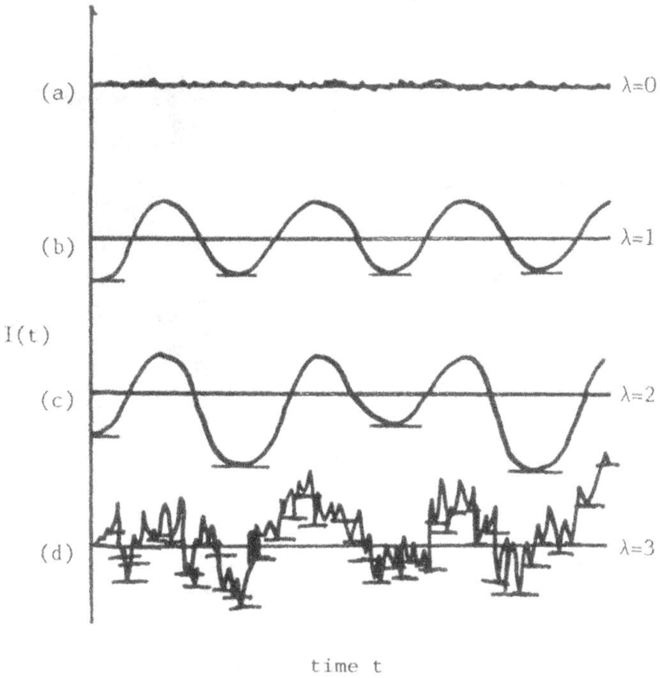

Figure 9. Real-time response for four values of the control parameter λ. Local minima are underlined.

drastically altered. Bifurcation diagrams provide a very useful visual display of the behavior of the system for various control parameters.

3. THE ANHARMONIC OSCILLATOR

Although the assumption of a harmonic potential is often a useful simplification in the description of real physical systems, that

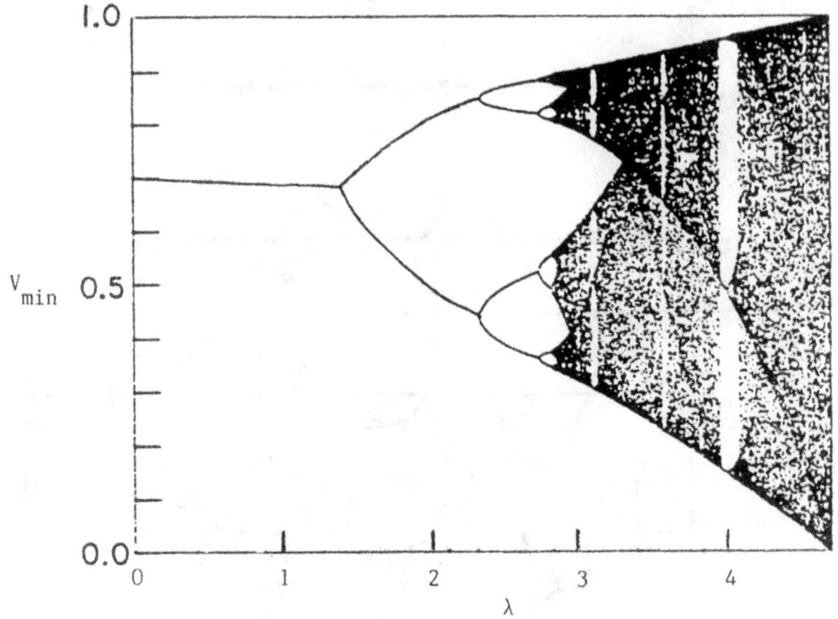

Figure 10. Bifurcation diagram.

simplification eliminates chaotic response! The anharmonic potential, on the other hand, is often more realistic, and usually leads to a rich spectrum of nonlinear response phenomena, including chaos. Huberman and Crutchfield [7] studied the solutions of a driven anharmonic oscillator nearly a decade ago, and identified the existence of a strange attractor and chaotic solutions.

We consider here the anharmonic restoring potential

$$U = \frac{1}{2}\phi^2 - \frac{1}{4}\phi^4 \tag{3.1}$$

or the related periodic potential

$$U = 1 - \cos\phi\ . \tag{3.2}$$

In certain approximations, the potential of Eq. (3.2) applies to two well-known solid state systems, Josephson junctions and charge density waves.

3.1 Josephson Junctions

A Josephson junction is an electronic device fabricated by sandwiching an insulating layer between two superconductors.[8] The resulting junction can be modelled by an ideal junction in parallel with a (quasiparticle) leakage resistance R and a junction capacitance C. The ideal junction obeys $I = I_c \sin\theta$ and $V = h\dot{\theta}/2e$, with I_c the junction current, V the junction voltage, and θ the superconducting phase difference across the junction. I_c is the critical current (beyond which a finite voltage appears across the junction). The resulting "RSJ" (for resistively shunted junction) equation of motion for the real junction becomes

$$\frac{\hbar C}{2e}\ddot{\theta} + \frac{h}{2eR}\ddot{\theta} + I_c \sin\theta = I\ , \tag{3.3}$$

with I the external drive current, $I = I_{dc} + I_{ac}\cos(\omega t)$. In appropriately redefined units (including time), Eq. (3.3) becomes

$$\beta\ddot{\phi} + \dot{\phi} + \sin\phi = I/I_c\ , \tag{3.4}$$

where β is a damping parameter. Equation (3.3) is isomorphic to that which describes a damped physical pendulum. In many cases it is assumed that

Eq. (3.4) describes so well the real junction that the detailed behavior of the junction and numerical integrations of the equation are taken as equivalent.

Equation (3.4) displays extremely complex nonlinear behavior,[9-13] and some of these features have recently been observed in real junctions. An important experimental difficulty with real junctions is, however, that the characteristic frequency is so high that only limited data acquisition and analysis is possible.

Figure 11 shows an early analog-computer determination[14] of the response of Eq. (3.4) for ac drive only, where $I = I_{ac} \cos(\omega t)$. At frequencies close to the natural oscillation frequency ω_o, an increasing ac amplitude leads to chaos. Later work[9-13] has shown that the situation is far more complicated than Fig. 11 would suggest, but the general result of chaotic solutions for ω close to ω_o and for sufficiently high ac drive still holds. Ac-induced chaos probably does exist in real Josephson junctions, but it has not yet been clearly observed.

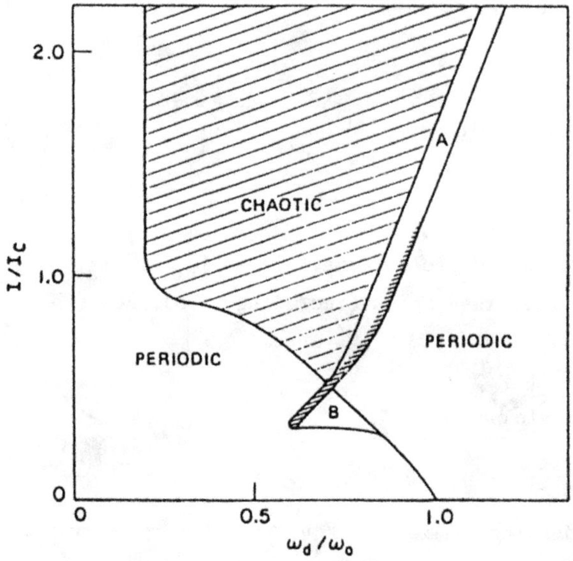

Figure 11. Response of Eq. (3.4) for $\beta^2 = 5$. From Ref. 14.

For combined ac and dc drive, i.e., for $I = I_{dc} + I_{ac}\cos(\omega t)$, Eq. (3.4) again predicts highly complex response. Mode-locked steps result whenever the external frequency matches the internal ac Josephson frequency.[8] Chaotic response can occur on the steps, causing excessive noise, or even leading to a destruction of the stability of the step.[9,13] Figure 12a shows such behavior in a Pb/Te/Pb Josephson junction.[15] Figure 12b shows the corresponding prediction of Eq. (3.4), which is seen to be in good agreement with experiment. Again, because of the very high frequencies involved in Josephson junction experiments (typically GHz), very little of the sophisticated analysis discussed in the previous section appears possible. Of course, a great deal of such analysis has been performed on Eq. (3.4), where the frequency is normalized to a convenient unit in the analog or digital computations.

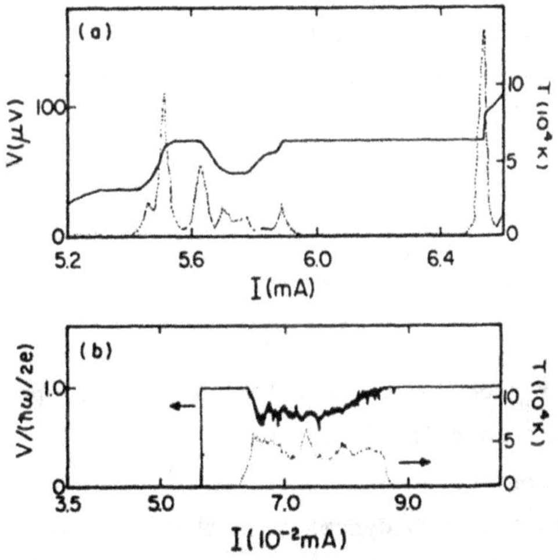

Figure 12. I-V characteristic (solid line) and noise temperature (dotted line) for Josephson junction driven by ac and dc current. (a): real Pb/Te/Pb junction; (b): theory. From Ref. 15.

3.2 Charge Density Waves

A charge density wave (CDW) is a modulation of the electronic charge density which results from a Peierls distortion in a solid state crystal. CDW transitions are a common phenomenon in a variety of low dimensional conductors.[16] The low dimensionality is important because good Fermi surface nesting helps drive the transition. In the quasi-1D conductor $(TaSe_4)_2I$, for example, there occurs a Peierls distortion at T_P = 265K; below T_P a collective-mode CDW state exists.

Electrons "condensed" in the CDW state respond to applied elecric fields in a coherent and unusual way. The phase of the CDW may be pinned to the underlying lattice by impurities, and only for dc fields $E_{dc} > E_T$, with E_T the threshold field, will the CDW electrons move and carry a current. Hence the CDW system is highly nonlinear. A moving CDW also generates an intrinsic narrow-band frequency, which is directly proportional to the CDW drift velocity. For low amplitude ac fields, the CDW responds much like an overdamped harmonic oscillator. In the simplest approximation, an equation of motion for the CDW phase may be written as[17]

$$\beta\ddot{\phi} + \dot{\phi} + \sin\phi = E/E_T, \tag{3.5}$$

where $E = E_{dc} + E_{ac}\cos(\omega t)$. Note that with appropriate redefinition of variables, this equation is identical to Eq. (3.4), which describes the dynamics of Josephson junctions. Equation (3.5) is admittedly a very crude approximation to CDW dynamics in that it fully neglects internal degrees of freedom for the CDW. Nevertheless, it does predict a threshold field E_T, generation of an intrinsic frequency linear in CDW drift velocity, and a harmonic-oscillator-type ac conductivity.

For sufficiently large ac driving fields, the CDW pinning potential is anharmonic, and one might therefore expect chaotic response for an ac driven CDW. In general, however, this is not observed (it has been searched for in $NbSe_3$, $(TaSe_4)_2I$, TaS , and $K_{0.3}MoO_3$, all sliding CDW systems). The problem lies in the excessive damping associated with CDW dynamics. If damping is too large, Eq. (3.5) does not predict chaos. This is illustrated in Fig. 13, which shows [12] the state diagram for Eq. (3.5) in the amplitude-damping plane (low β corresponds to large damping; $\rho \propto E_{ac}$). For large damping, a very large ac amplitude is required for non-periodic response.

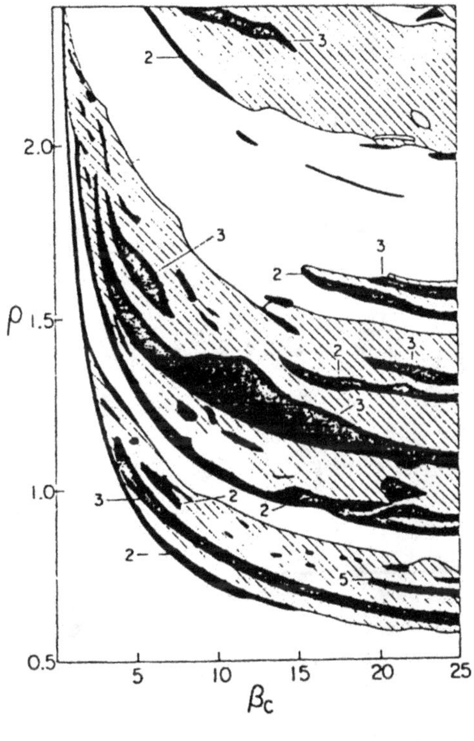

Figure 13. State diagram of Eq. (3.5) in the drive amplitude-damping plane. The hatched regions are chaotic. From Ref. 12.

In $(TaSe_4)_2I$, the overall damping can be modified by placing the CDW crystal in series with an inductance.[18] The inductance simulates CDW inertia. The resulting hybrid circuit does display chaos, as illustrated in Fig. 14. The general form of the response should be compared to Fig. 11, which is the prediction of Eq. (3.5) with small damping.

CDW systems offer distinct advantages over Josephson junctions in the study of chaos, in that the characteristic CDW frequencies are in

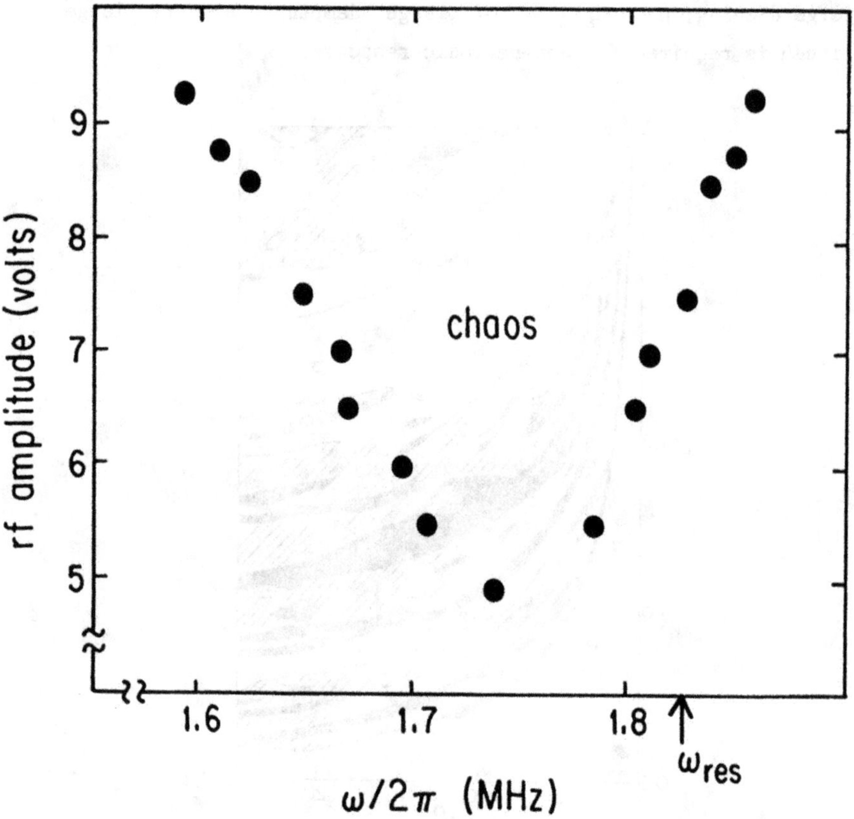

Figure 14. Chaotic-periodic response boundary for a $(TaSe_4)_2I$ circuit. From ref. 18.

the low MHz range. In sharp contrast to the GHz range of Josephson junctions, a host of rf electronics is available for detailed MHz chaos analysis.

4. MODE LOCKING

Mode locking is a rather general phenomenon which can occur in systems with competing periodicities. Let us consider a system with an internal periodicity or frequency ω_{in}, driven externally by a source at frequency ω_{ex}. Depending on the coupling between the external drive and the system, the internal frequency can "lock" to ω_{ex}. This is the process involved in biological pacemakers, where the (otherwise erratic) internal frequency of the heart is locked to the carefully selected drive frequency of an electronic pulse generator.

We shall here discuss mode locking in the language of the circle map, which provides a concise yet far reaching description of chaotic behavior in a variety of real systems.

4.1 Circle Map

Imagine the quasi-periodic behavior of a system with an internal frequency $\omega_{in} \sim \dot{\Theta}_1$ driven by an incommensurate source at frequency $\omega_{ex} \sim \dot{\Theta}_2$. The motion of the system in phase space can be represented as motion of a 2-D torus, as illustrated in Fig. 15. We may select a Poincaré section at the plane $\Theta_2 = 0$, and strobe the system at "times" $\Theta_2 = 0,\ 2\pi,\ 4\pi$, etc. As the system evolves, each intersection point on

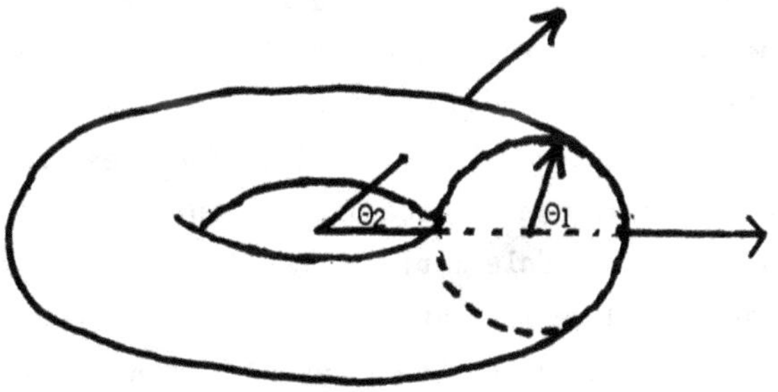

Figure 15. Motion on a 2-D torus. The intersection with the plane $\Theta_2 = 0$ defines the circle (or topological equivalent).

the circle (the circle is defined by the intersection of the torus with the plane) is mapped to another point on the same (invariant) circle. We thus have a nonlinear map of the circle onto itself .

One such mapping of the invarient circle is the sine circle map,[19]

$$\theta_{n+1} = \theta_n + \Omega + \frac{K}{2\pi} \sin 2\pi\theta_n , \tag{4.1}$$

where θ_n describes the state of the system at index n, $\Omega = \omega_{in}/\omega_{ex}$, and K is the "strength" of the coupling between the system and the external drive. The sine circle map has been extensively studied, and predicts transitions to chaos.[19-22]

Figure 16 shows numerous iterations of the sine circle map with $\Omega =$ 0.2 and K = 0.9 .[21] Quasi - periodic behavior is observed. With increasing K, the map develops a local maximum; hence it is no longer invertible and may develop chaos. For certain values of K and Ω, the circle map predicts <u>mode locking</u>, where ω_{in} is related to ω_{ex} by $\omega_{in}/\omega_{ex} = p/q$, with p and q integers. Figure 17 shows some of the mode-locked regions of the circle map in K-Ω space.[21] The locked regions are often referred to as "Arnold tongues" or "entrainment horns". For large K, the tongues or resonances overlap, and this overlap can lead to hysteresis and chaos (hopping between resonances). If all p/q ratios had been plotted in Fig. 17, the resonances would start overlapping at K = 1, the critical line. Only for K > 1 is chaos possible in the sine circle map.

The mode-locked regions at K = 1 in the circle map form a Cantor set [23] with a fractal dimension. Figure 18 shows the set of locked regions; the set forms a "<u>devil's staircase</u>". The dimension of the (complete) staircase at K = 1 has been computed as $d_F = 0.860$.[20-22]

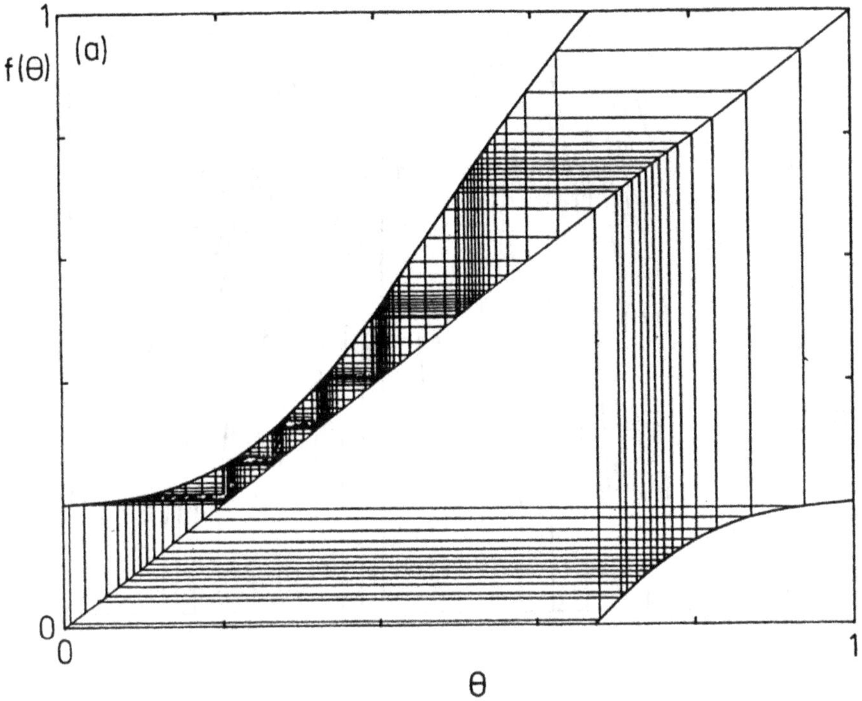

Figure 16. Iterating the sine circle map in the quasi-periodic regime ($K = 0.9$, $\Omega = 0.2$). From Ref. 21.

It has been clearly demonstrated that the sine circle map does apply to a number of real physical systems, in particular fluid instabilities [24] and some semiconductors.[25] We now consider its potential applicability to Josephson junctions and charge density waves.

4.2 Josephson Junctions and Charge Density Waves

Mode locking phenomena occur in Josephson junctions and are well known in the Josephson literature.[8] In general the experiment is performed by irradiating the junction with microwaves at frequency ω_{ex}, and measuring the dc I-V characteristics. The locked regions are

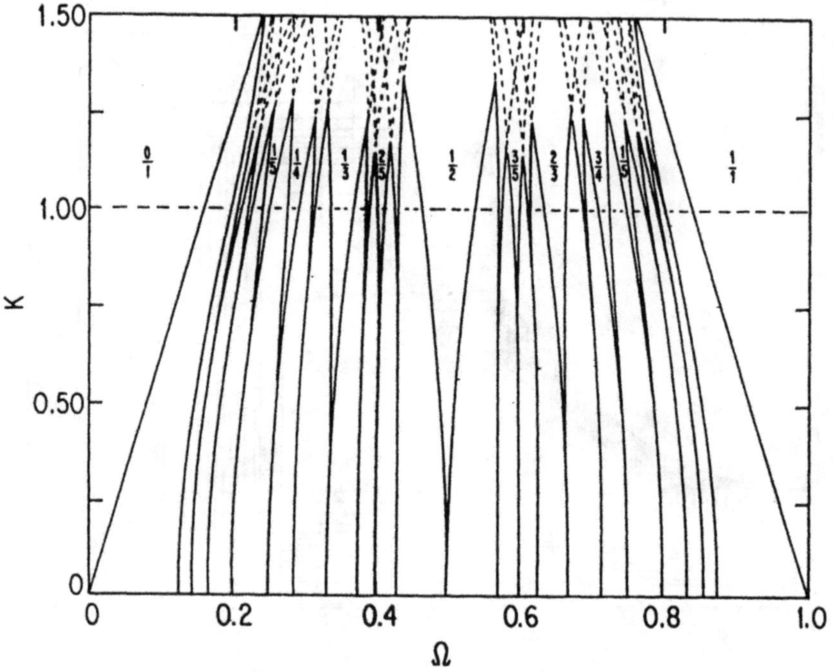

Figure 17. Mode-locked regions of the sine circle map. The corresponding ratios p/q are indicated. From Ref. 21.

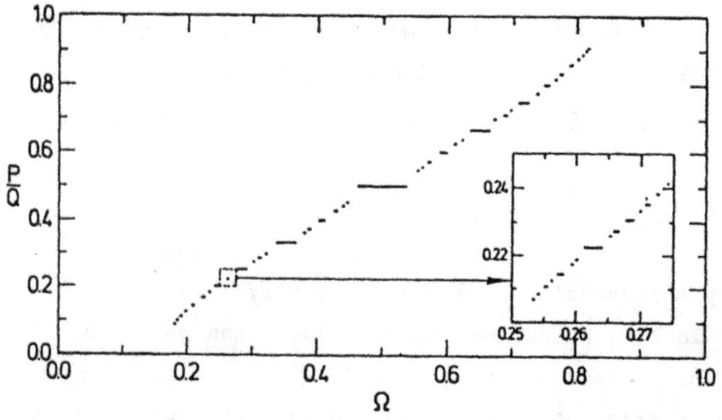

Figure 18. Devil's staircase. The structure is self-similar. From Ref. 20.

indicated by steps in the I-V curve. Analog computer solutions [26] of the Josephson equation, Eq. (3.3), have revealed that at K = 1, a mode locked interval with fractal dimension d_F = 0.87 is obtained, as suggested by the circle map. Chaos (as evidenced by hysteresis and noise in the I-V characteristics) has also been demonstrated in real junctions. [27]

In the case of charge density waves, the situation is quite complicated. For combined ac and dc electric drive fields, mode-locking is indeed observed, as illustrated in Fig. 19. [28] The flat-top peaks reflect the CDW phase being locked such that the CDW drift velocity is dictated by $v_d = \lambda_{CDW}\omega_{ex}p/q$, i.e., $\omega_{in}/\omega_{ex} = p/q$ over a finite range of E_{dc}. λ_{CDW} is the CDW wavelength. From data similar

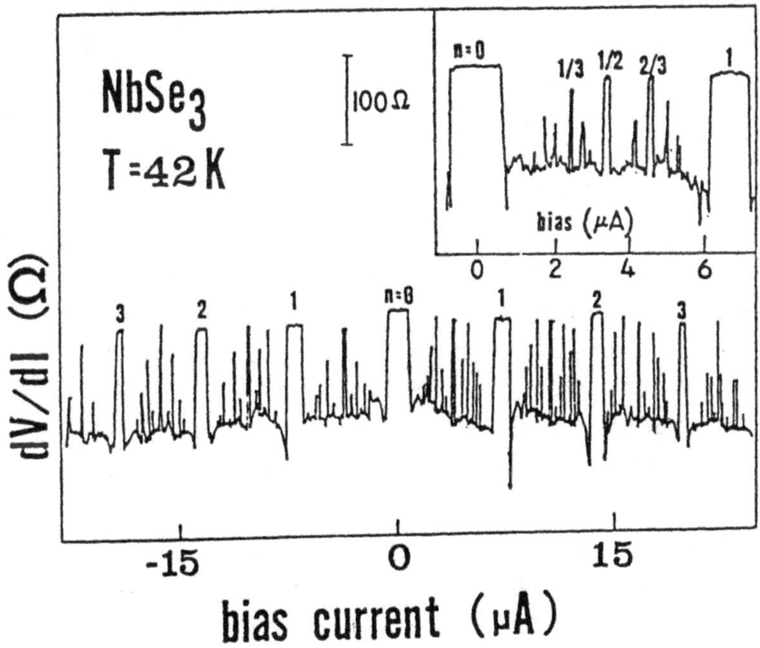

Figure 19. Mode locking in the CDW system $NbSe_3$. From Ref. 28.

to that of Fig. 19, the fractal dimension of the complementary Cantor set of the mode-locked region can be determined. This is performed as follows:[20-22] given a scale r, the total measure of the gaps between the mode locked steps is $1 - S(r)$, and the number of "holes" is $N(r) = [1 - S(r)]/r$. If $rN(r) \longrightarrow 0$ as $r \longrightarrow 0$, the staircase is complete, with fractal dimension given by

$$N(r) \sim r^{-d_F} . \tag{4.2}$$

An early analysis of $NbSe_3$ data by Brown et al.[29] yielded $d_F = 0.91$. More recent studies [30] find that d_F depends strongly on ac drive magnitude, and actually approaches 1 as E_{ac} is increased. Chaos is not observed. This behavior is not entirely consistent with the circle map, and suggests that internal CDW degrees of freedom play an important role in the mode locking phenomenon.[31]

Figure 20a shows mode-locked regions (including harmonic and subharmonic structure) computed from coupled overdamped oscillators.[32] It is important to note that Eq. (3.5) does not predict subharmonic structure (nonintegral p/q ratio) in the overdamped limit. The behavior of Fig. 20a appears to be in good agreement with that observed in real CDW systems. Hence, the CDW may be more appropriately described as a set of coupled nonlinear oscillators, rather than a single anharmonic oscillator. As shown in Fig. 20b, even the set of coupled overdamped oscillators does not display any positive Lyapunov exponents, and thus no chaos.

The CDW systems discussed so far are overdamped and do not in themselves display chaotic response. The situation is very different for CDW systems in the "switching" regime. In switching CDW's, phase slip centers are present in the crystal which dramatically alter the nonlinear dynamics.[33]

Figure 21a shows mode-locked steps for $NbSe_3$ in the switching regime. On each mode-locked step, a period doubling route to chaos is

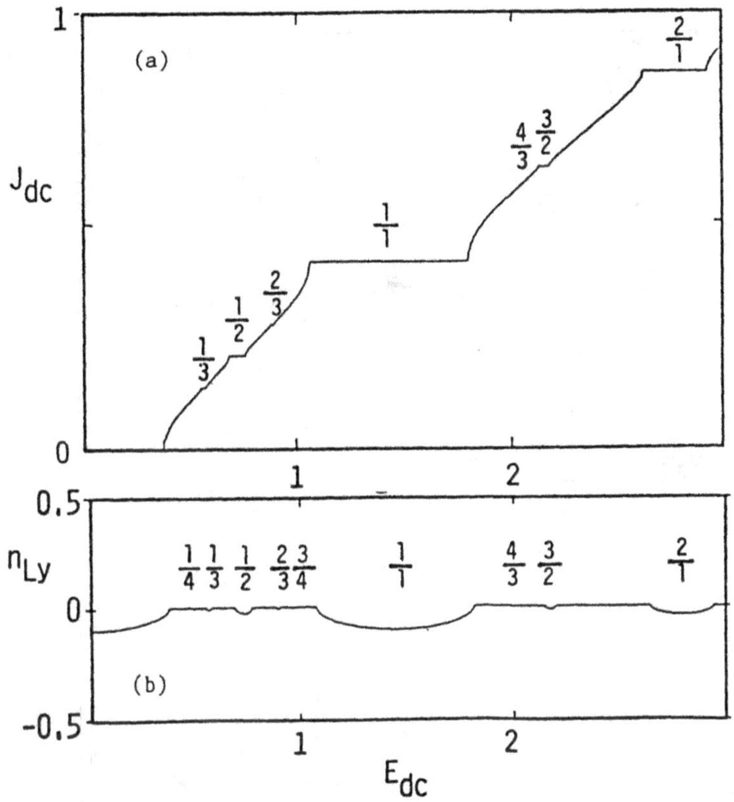

Figure 20. Locking in a system of two coupled oscillators. a) steps in I-V characteristics; b) Lyapunov exponents in mode-locked regions. No chaos is observed. From Ref. 32.

observed, as evidenced by the corresponding power spectra displayed in Fig. 21b. An interesting feature of the period doubling route is that it is observed to fully repeat on each new step; this behavior is consistent with the predictions of the circle map. In switching samples at large ac drive the fractal dimension of the mode locked steps approaches $d_F = 0.85$.[30] The routes to chaos in CDW systems, in

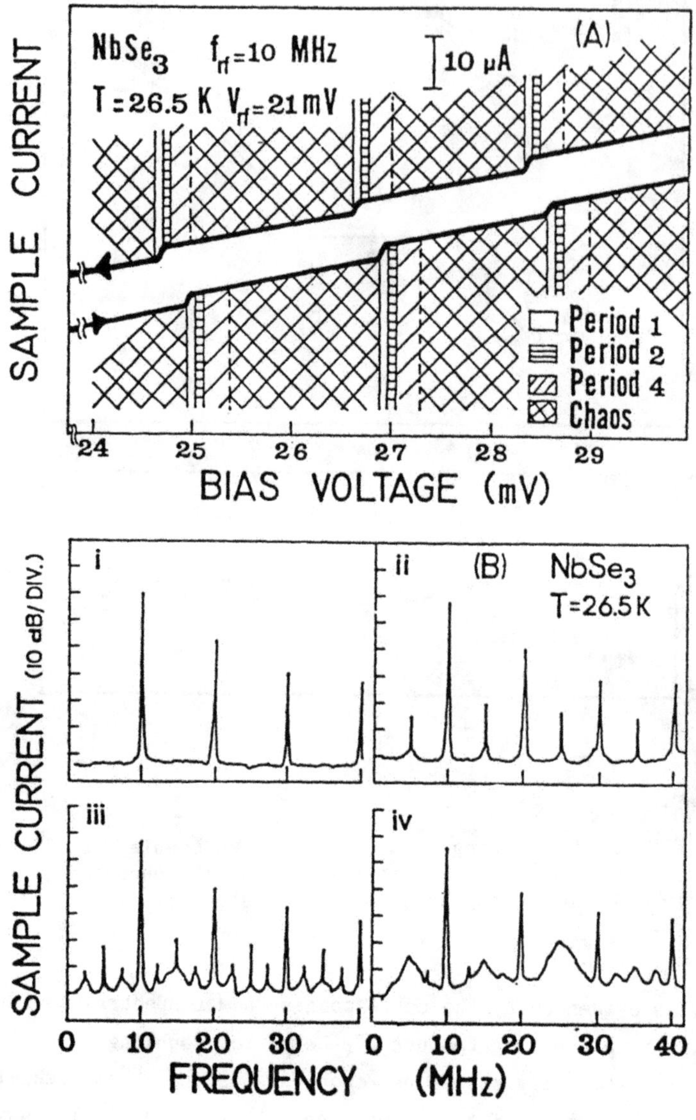

Figure 21. Period doubling route to chaos in $NbSe_3$. From Ref. 30.

particular dynamics in the switching regime, are largely unexplored, and should provide a bountiful area for future chaos research.

5. SEMICONDUCTORS

A great deal of very careful and beautiful work on nonlinear dynamics has recently been performed on semiconductor devices, including p-n junctions, photoconductors, and electron hole plasmas. The great precision with which the data can be collected allows detailed analysis, and hence these experiments have provided stringent tests of nonlinear dynamics theory.

Let us examine some "universal" predictions of nonlinear dynamics theory. The treatment is restricted to the period doubling route to chaos for a 1-D map with a single quadratic maximum. As a control parameter λ is increased, successive bifurcations occur, and eventually chaos is achieved. This is illustrated schematically in Fig. 22.

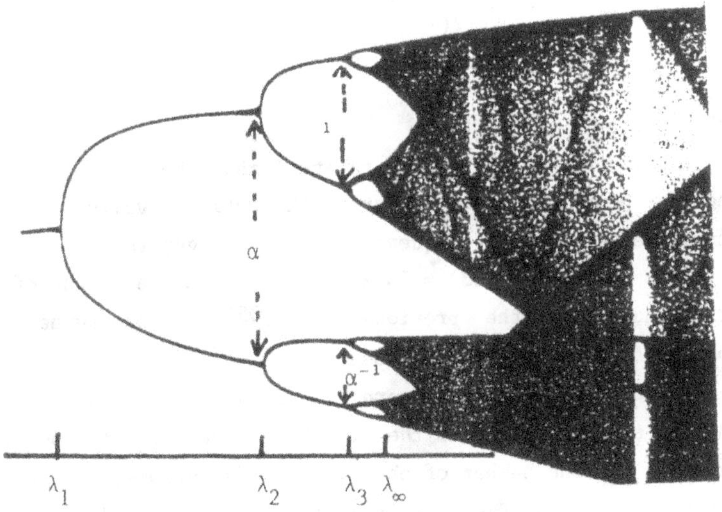

Figure 22. Bifurcation diagram.

Feigenbaum [34] has predicted that there exists a well-defined scaling of the control parameter evaluated at the bifurcation points. At the first bifurcation, $\lambda = \lambda_1$, at the second $\lambda = \lambda_2$, and so on. At $\lambda = \lambda_\infty$, the system goes aperiodic. It is suggested [34] that λ_n converges to λ_∞ geometrically, i.e.,

$$(\lambda_\infty - \lambda_n) \sim \delta^{-n} , \tag{5.1}$$

valid for large n, or that $\delta_n = (\lambda_{n+1} - \lambda_n)/(\lambda_{n+2} - \lambda_{n+1})$ converges to $\delta \simeq 4.669$ as n becomes large. This value for δ is a "universal" number, in that it should not depend on the details of the system.

A second universal number relates the vertical separation d of the bifurcation points on a bifurcation diagram, as illustrated also in Fig. 22. For large n, scaling exists [34] such that

$$\frac{d_{n+1}}{d_{n+2(+)}} = \alpha \; ; \quad \frac{d_{n+1}}{d_{n+2(-)}} = \alpha^2 , \tag{5.2}$$

with $\alpha \simeq 2.502$.

A third universal number relates that heights (power) of new frequency peaks which emerge during the period doubling sequence. As successive subharmonic frequency peaks are generated, their average height is 10 log (20.963) = 13.21 db down from the height of the peak corresponding to the previous point.[35] The number 13.21 db represents a universal power loss.

We shall consider two more universal behaviors in the period doubling sequence. One is the amount of external noise power needed to reduce by one the number of observable bifurcations; this turns out to be a noise increase factor of $N = 6.55$.[36] The second relates to noise free "windows" in the bifurcation diagram as the control parameter is increased. For a 1-D map with a single extremum (not necessarily a quadratic), a universal "U" sequence is followed.[37)] Chaotic regions along λ-space are interrupted by regions of periodic behavior with no noise. Counting periods less than or equal to 6, the U-sequence is 1,2,4,6,5,3,6,5,6,4,6,5,6,..., where the numbers correspond to the period of the successive noise free windows.

The above features have been confirmed in numerical studies of various mappings. In the following section we shall test these features in a specific solid state system, the p-n junction.

5.1 P-N Junction

A p-n junction (diode) is a highly nonlinear device. When coupled to an external series resistance and inductance, the system can be driven into a chaotic state via a period doubling route.[38)] Figure 23 shows a bifurcation diagram for a particular diode model. If the control parameter value at the bifurcation points is scaled according

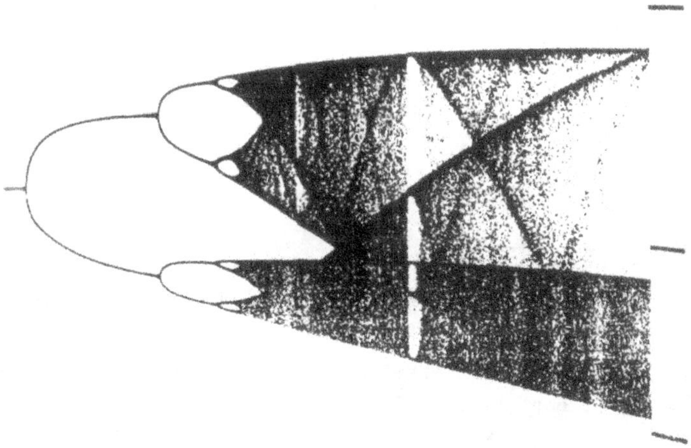

Figure 23. Bifurcation diagram for diode #300A. From Ref. 38.

to Eq. (5.1), δ may be obtained. Similarly, α, N, and the power down ratio of the new frequency peaks can easily be determined. Table 5.1 summarizes some of the experimental and theoretical results for these parameters.[38] The experimental results are in surprisingly good agreement with theory (considering, for example, the limited range of n available experimentally). The periodicity of noise free windows in the bifurcation diagram has also been determined for various diodes. An observed sequence is 6,5,7,3,6,12,9,13, which is in exact agreement with the U-sequence, starting at the first 6 (see above).

Number		Measured	Predicted
δ_1	Eq. (3)	4.26 ± 0.1	4.751
δ_2	Eq. (3)	4.28 ± 0.1	4.656
δ_1	Period 3 window	0.69 ± 0.1	0.979
δ_2	Period 3 window	3.38 ± 0.1	4.429
α		2.41 ± 0.1	2.502
N		6.3 ± 0.3	6.55
Average spectral power ratio		11 to 15 dB	13.61 dB

Table 5.1. Universal numbers, as measured from p-n junctions and predicted from theory. From Ref. 38.

5.2 Coupled Oscillators

If a number of identical p-n junctions are coupled together and driven from a single source, the resulting system may exhibit an exceedingly complex response spectrum.[39] Of particular interest is the possibility of Hopf bifurcations, which correspond to the birth of a limit cycle. Figure 24 shows schematically a Hopf bifurcation as a control parameter λ is varied. At the bifurcation point a new (incommensurate) periodicity is exhibited. In practice, this results in a new, seemingly unrelated, frequency peak appearing in the response power spectrum.

Van Buskirk and Jeffries [39] have extensively studied the dynamics

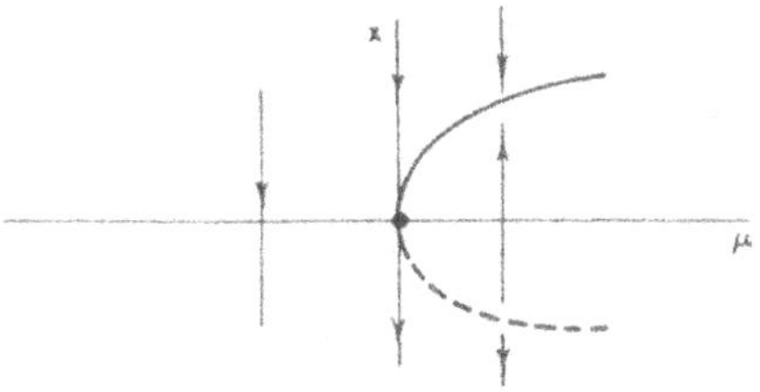

Saddle-node bifurcation.

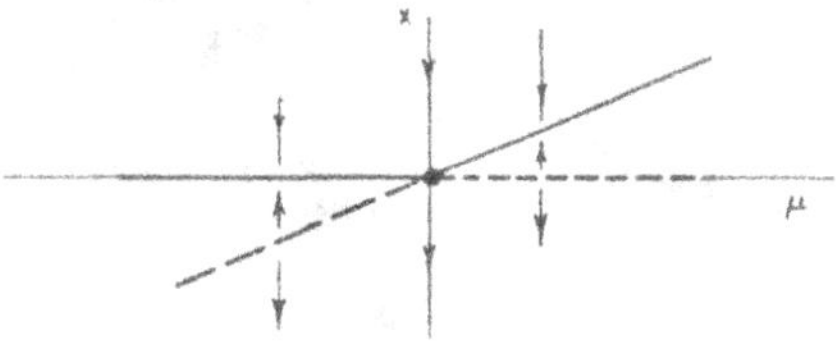

Transcritical bifurcation.

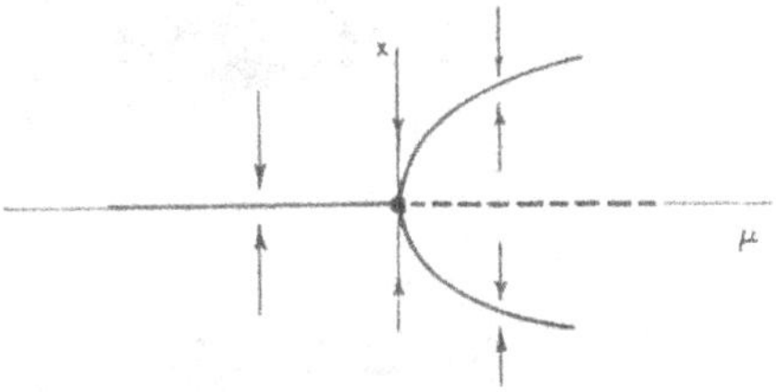

Pitchfork bifurcation (supercritical).

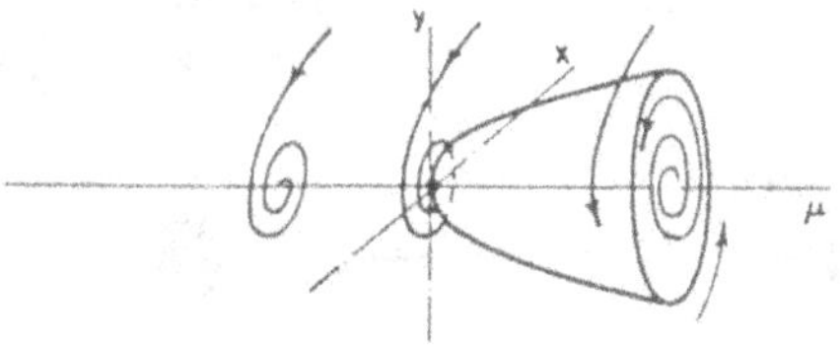

Hopf bifurcation (supercritical).

Figure 24. Co-dimension one bifurcations. The Hopf bifurcation corresponds to the birth of a limit cycle. Adapted from J. Guckenheimer and P. Holmes, Nonlinear Oscillations, Dynamical Systems, and Bifurcation of Vector Fields (Springer, New York, 1983).

of two coupled p-n junctions. For "resistively coupled" junctions, the bifurcation diagram displayed in Fig. 25 is obtained. For this system the initial period doubling is followed by a Hopf bifurcation (the additional incommensurate frequency "fills up" the area between the diagram boundaries). This is followed by mode locking (or

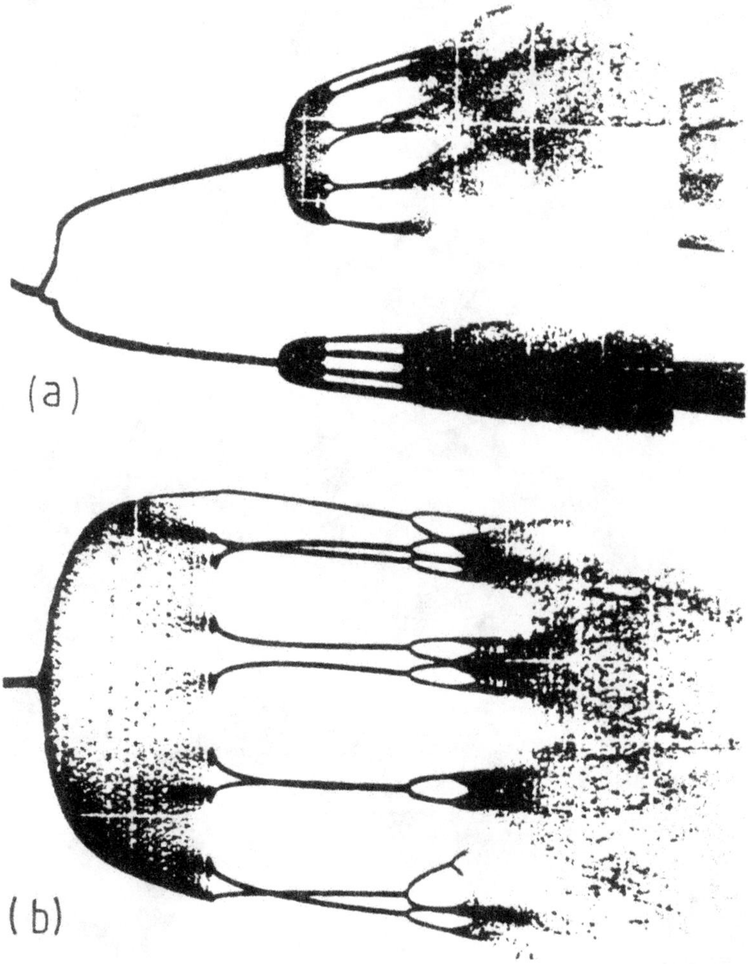

Figure 25. Bifurcation diagram for two coupled oscillators. From Ref. 39.

entrainment), two-band chaos, and finally crisis [40] of the attractor, i.e., a drastic change in the topological nature of the attractor. From the experimental data a "phase diagram" for the response of coupled p-n junctions can be constructed, as shown in Fig. 26. Note the complexity of this phase diagram when compared to the relatively simple circle map predictions of Fig. 17. A number of very beautiful data sets (phase

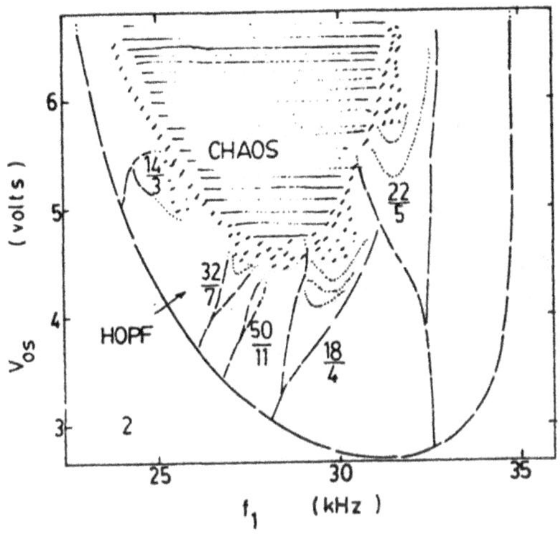

Figure 26. Phase diagram for coupled p-n junction oscillators. From Ref. 39.

portraits, Poincaré sections, bifurcation diagrams, etc.) have been obtained for coupled p-n junctions, and are described in detail in Reference 39. In this reference the p-n junctions are also modelled by various differential equations and return maps.

It should be noted that coupled p-n junctions bear a certain resemblance to some charge density wave compounds, where individual "oscillating" domains may be coupled together by normal electron resistive shunts. Phenomena such as Hopf bifurcations have yet to be explored in CDW systems.

5.3 Photoconductors

Helium-cooled extrinsic photoconductors are often used as high-sensitivity, low-noise detectors of far infrared radiation. For this application, chaotic behavior is highly undesirable, but it does exist. From a nonlinear dynamics point of view, photoconductors are thus of interest. Chaos in photoconductors was first identified by Teitsworth et al.,[41] who found a period doubling route to chaos induced by an applied dc field. Similar routes to chaos have also been obtained for ac drive.

Figure 27 shows a phase portrait and corresponding power spectrum for an ac-driven Ge photoconductor, at various values of ac drive amplitude.[42] Both the phase plot and power spectrum indicate successive period doubling with increasing ac drive. In Fig. 28, the regions of different response are mapped out in the phase space of drive amplitude and drive frequency. The results of Fig. 27 and 28 are well described by a simple and standard rate equation model for the photoconductor,[43]

$$\dot{p} = \gamma(a - a_*) + pK(a - a_*) - pra_* , \tag{5.3a}$$

$$\varepsilon E = J_{ext} - epv , \tag{5.3b}$$

$$p = a_* - d , \tag{5.3c}$$

$$J_{ext}(t) = J_0 + \Delta J \sin(\omega_{ex} t), \tag{5.3d}$$

where p is the hole concentration in the photoconductor, a is the total acceptor concentration, γ is the generation rate, K and r denote the

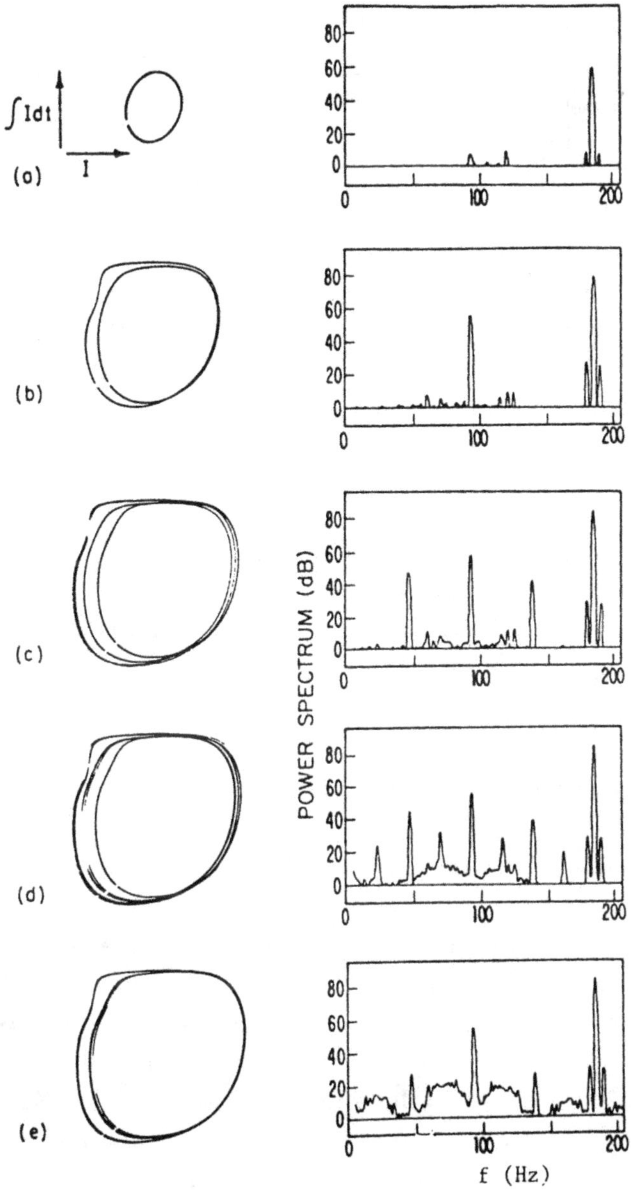

Figure 27. Response of Ge photoconductor for increasing ac drive amplitude. From Ref. 42.

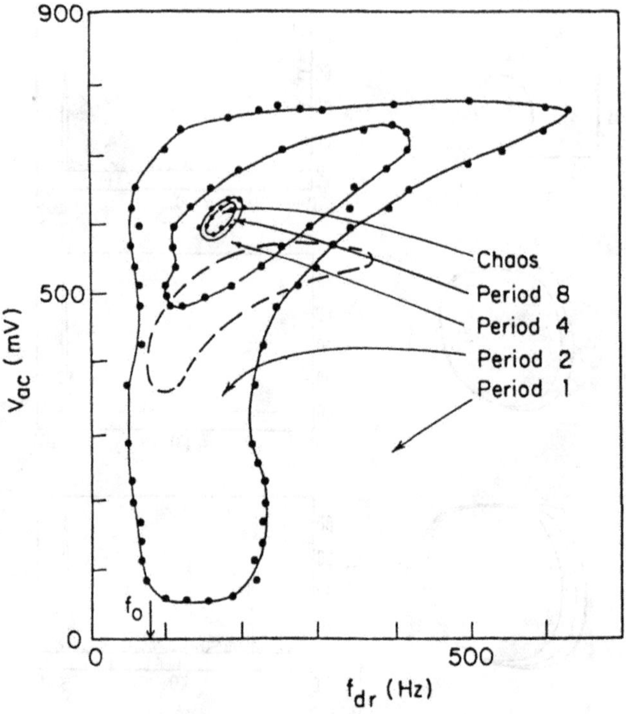

Figure 28. Response phase plot for Ge photoconductor. From Ref. 42.

field dependent impact ionization and capture rates, respectively, J_{ext} is the external current density, v is the carrier drift velocity, and ΔJ is the amplitude of the induced modulation current density. Numerical solution [43] of these equations produces the response displayed in Fig. 29. Figure 30 shows the corresponding Poincaré section and return map, and Fig. 31 shows the regions of periodic and chaotic response in the space of ac drive amplitude and drive frequency. Figure 31 may be qualitatively compared to the experimental data of Fig. 28. It is rather surprising that the set of simple standard equations (5.3) yield so complex a response, and in good qualitative agreement with experiment.

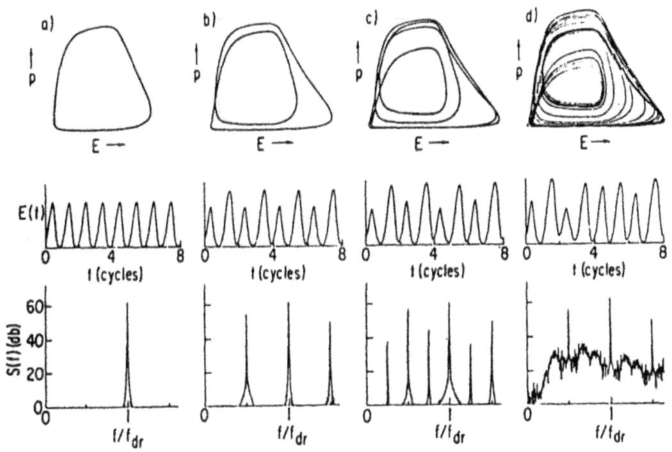

Figure 29. Phase portrait, real-time output, and power spectrum for photoconductor model, Eq. (5.3). From Ref. 43.

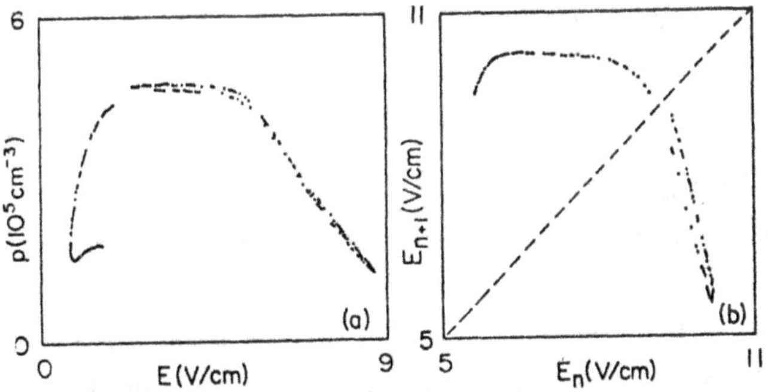

Figure 30. Poincaré section and return map generated from photoconductor model, Eq. (5.3). From Ref. 43.

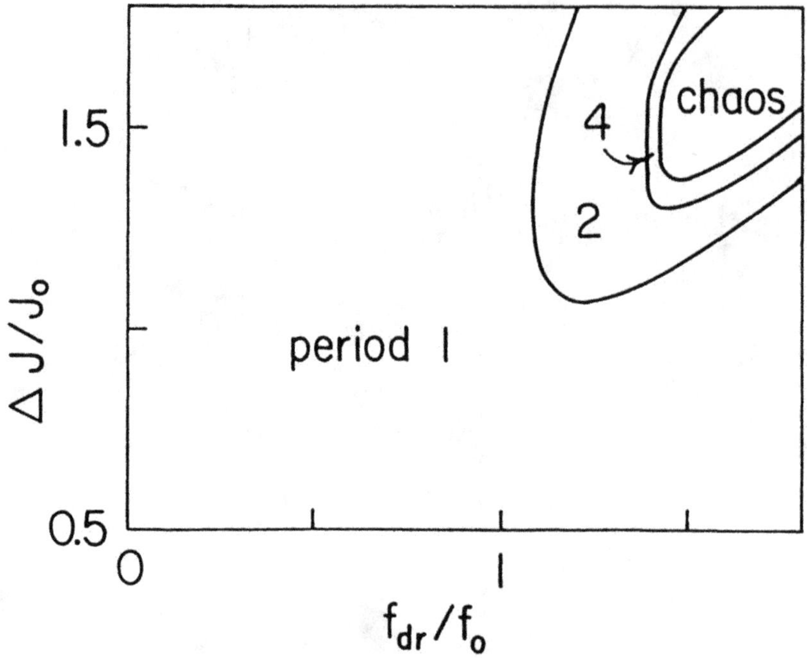

Figure 31. Response boundaries predicted by photoconductor model, Eq. (5.3). From Ref. 43.

5.4 Electron-Hole Plasma

The electron-hole (e-h) plasma in Ge represents probably the first solid state system for which chaotic response has been studied spatially as well as temporally. We consider first the e-h plasma in a cylindrical bar of semiconductor, with a long axis in the z direction, as illustrated in Fig. 32.[44] With applied parallel E_{dc} and H fields in the z direction, a screw-shaped helical wave is generated within the plasma, corresponding to a region of enhanced electron and hole density. The helical density is not stable, but may travel, thereby generating internal oscillations within the specimen.[44]

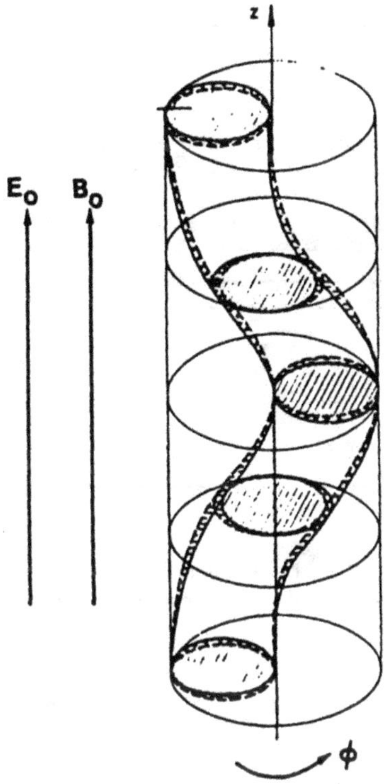

Figure 32. Helical plasma density in Ge. The electric and magnetic fields are along the axis of the specimen. From Ref. 44.

Various routes to chaos in the e-h plasma of Ge have been studied experimentally by Held and Jeffries,[45,46] using the geometry of Fig. 33. For the study of temporal chaos alone, the dc drive voltage V_{dc} is applied to the ends (injection contacts) of the bar. The system response is the current through the specimen. Both period doubling and quasi-periodic routes are observed.

Figure 34 shows an example of the period doubling route in the e-h plasma in Ge, with H = 4kG. The current response in real time, phase

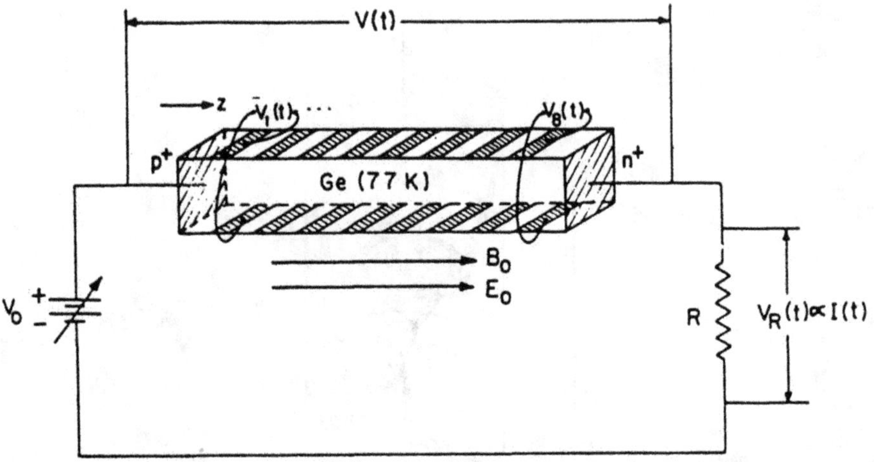

Figure 33. Experimental configuration for chaos measurements on electron hole plasma in Ge. From Ref. 46.

plots, and power spectra are shown. Fig. 34d corresponds to chaos, while Fig. 34e corresponds to a period 3 window. The quasi-periodic route to chaos is illustrated in Fig. 35. Here the experimentally determined return maps $\{I_{n+1}, I_n\}$ and power spectra are shown. Onset of chaos is observed in Fig. 35d; "strong" chaos is apparent in Fig. 35f.

The fractal dimension of the attractor associated with chaotic dynamics in the e-h plasma has been computed experimentally, using the method outlined in Sec. 2.2. Some uncertainty arises because not all randomly chosen points from the reconstructed phase space yield the same fractal dimension. Figure 36 shows a histogram of fractal dimension obtained from 27 independent determinations of the fractal dimension for identical drive conditions.[46)] The peak in the distribution suggests that the fractal dimension d_F lies between 2.4 and 2.6.

In Fig. 33, the Ge bar shown has a number of independent sensing probes along its length. These probes allow the investigation of spatial chaos, i.e., the response can be spatially correlated. If the

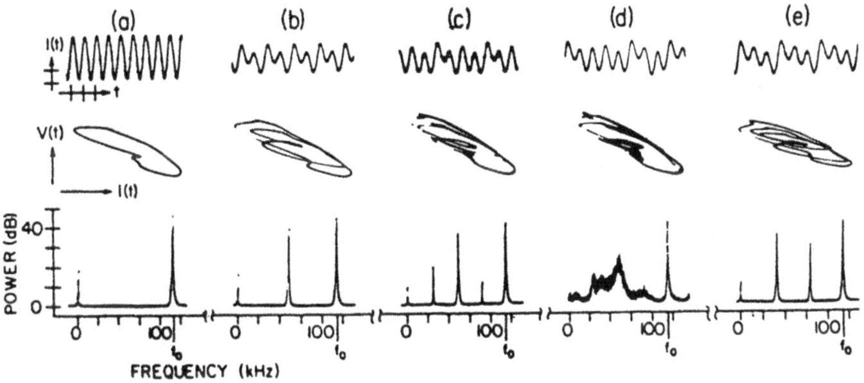

Figure 34. Period doubling route to chaos in e-h plasma in Ge. Real-time response, phase portraits, and power spectra are shown. The control parameter is the applied dc voltage. From Ref. 46.

temporal response were spatially coherent (regardless of the nature of the temporal response), then spatial correlation of the response would show only a phase shift, due to the coherent travelling e-h wave. The degree of spatial coherence is determined from a spatial correlation function [46)]

$$C(r) = \left| \frac{2}{N} \sum_{n=1}^{N} V_i(n\tau)\, V_j(n\tau) \right|^{1/2} . \qquad (5.4\ )$$

where $V_i(t)$ and $V_j(t)$ are the response voltages across two pairs of contacts separated by a distance r, t is the sampling interval, and N is a number large enough that C(r) has converged (typically, in these experiments, N = 20,000).

In the periodic regime of the driven e-h plasma, C(r) is that expected from a coherent travelling wave. However, in the (temporally) chaotic regime, there appears to be spatial breakup as well. Figure 37 shows C(r) versus V_{dc} for the e-h polasma for r = 4mm and 7mm. [46)] At V_{dc} = 22V, C(r) has dropped well below that expected for spatiallly

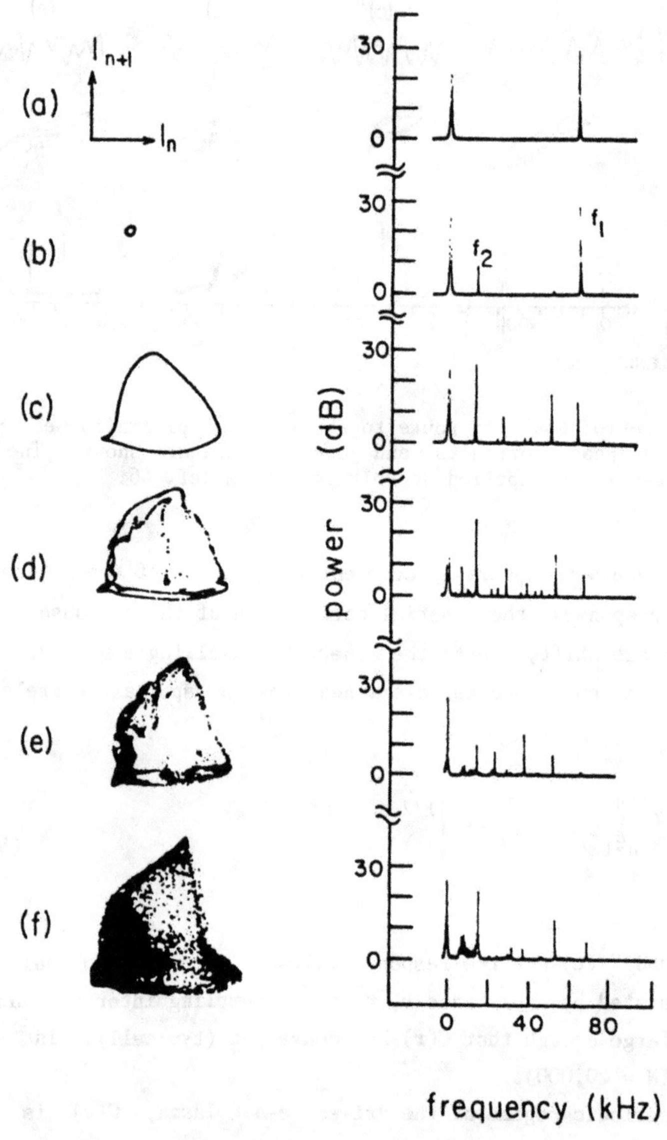

Figure 35. Experimentally determined return maps and power spectra for e-h plasma in Ge. A quasi-periodic route to chaos is indicated. From Ref. 46.

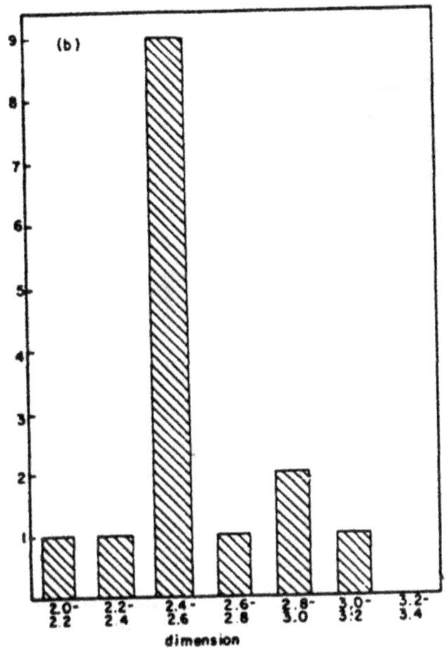

Figure 36. Dimension histogram for chaotic response in e-h plasma in Ge. From Ref. 46.

coherent response; this provides strong evidence for spatial chaos.

In the spatially chaotic regime of Ge, attempts have been made to determine the dimension of the attractor. Held [46] finds that just prior to breakup of the spatial coherence, the total current I(t) of the system is characterized by a low-dimensional attractor of dimension $d_F = 2.5$. Just after the onset of spatial disorder, dF is dramatically increased, with a lower limit of $d_F > 8$. The data for the determination of d_F in both cases is displayed in Fig. 38.

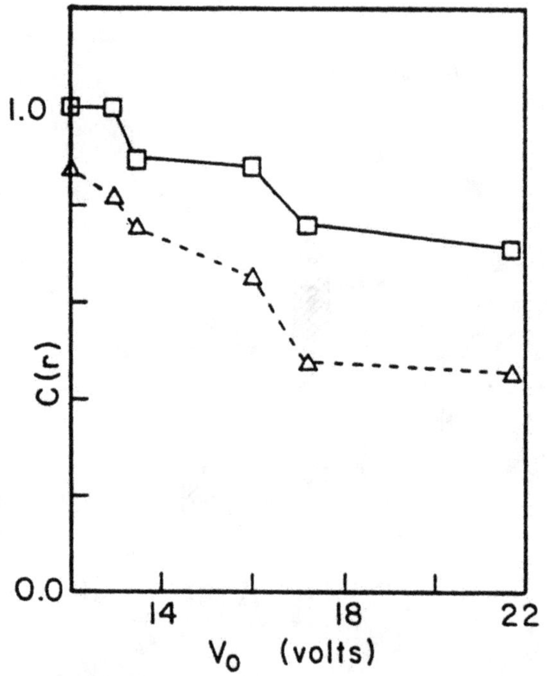

Figure 37. Spatial correlation versus dc drive in e-h plasma in Ge. From Ref. 46.

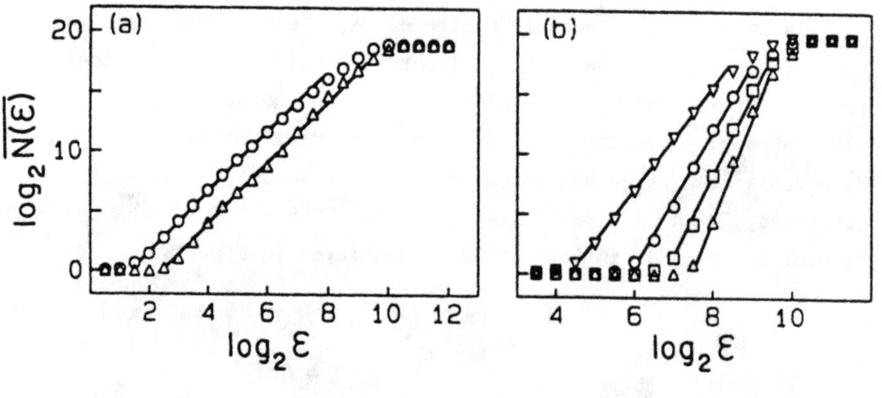

Figure 38. Fractal dimension determination for e-h plasma in Ge. a) in spatially coherent regime; b) in spatially turbulent regime. From Ref. 46.

The existence of spatial as well as temporal chaos no doubt exists in other solid state systems, such as magnetic chains and charge density waves. Indeed, recent experiments of the CDW system $NbSe_3$ have demonstrated that the source of the chaos (phase slip centers) corresponds to highly localized entities. [33)]

6. NOISE

One of the most notable features of chaos is of course the associated "noisy" response. In this section, we are interested not only in the noise generated by the chaos, but what influence external noise has on the system, in particular in the vicinity of a critical point. In Sec. 5, we have already encountered one external noise phenomenon: the amount of external noise needed to suppress the observation of one period doubling bifurcation. External noise (unrelated to noise generation by the chaos itself) can complicate noise analysis. For example, we have seen in the period doubling routes that a scaling relation exists for the parameter λ at bifurcation points. If noise is present, it may be difficult to determine exactly when the bifurcation takes place. One method of dealing with this problem is to asume that the period doubling peak amplitude in the Fourier transform scales as $(\lambda - \lambda_n)^{1/2}$, hence the peak height in the power spectrum scales as $\lambda - \lambda_n$. In practice one then needs only to measure the peak height over a small range of λ near λ_n, and extrapolate to zero height for λ_n.

In practice, the onset of a bifurcation (neglecting amplifier noise) is not always "clean", i.e. there often appears some sort of noisy precursor effect. In fact, the precursor effect can be predicted from chaos theory.

6.1 Noisy Precursor

Noisy precursor effects were first considered in detail by

Wiesenfeld.[47] We consider the effect of external Gaussian white noise of a nonlinear dynamical system. The noise acts as a stochastic perturbation on the forcing term, i.e., the limit cycle is not perturbed, and the phase space portaraits are "frozen in". Note that this approach is distinctly different from letting the parameters of the system fluctuate, which would lead to a fluctuation of the phase space portraits.

We here restrict ourselves to bifurcations of periodic orbits. In general we may write the perturbations in terms of eigenvectors

$$\phi^K(t) = e^{\rho_K t} \chi^K(t) , \qquad (6.1)$$

where the exponential term is the <u>Floquet multiplier</u> and the χ term is periodic. If the limit cycle is stable, then we must have $\mathrm{Re}(\rho_K) < 0$, i.e., the exponential term decays in time. There exist four kinds of bifurcations of periodic orbits: 1) transcritical ($\mathrm{Im}\rho_K = 0$); 2) symmetry breaking ($\mathrm{Im}\rho_K = 0$);3)Hopf ($\rho_K = e + ib$); and 4) period doubling ($\mathrm{Im}\rho_K = 1/2$). We shall consider the period doubling and Hopf cases.

Figure 39 shows a stable period T orbit X_0 in phase space, along with the Poincaré section (single point). If the control parameter λ were increased, this orbit might undergo a period doubling bifurcation at λ_1. However, let us keep λ near to, but less than λ_1. If we now perturb the system, it will relax back. Successive intersections with the Poincaré plane will yield the Poincaré section displayed in Fig. 40. The points alternate back and forth about the limiting point. These transients have the character of a damped, <u>period 2T</u> orbit; hence the associated noise spectrum will show a broad bump at one half the fundamental frequency of X_0. Figure 41 shows the expected behavior for the power spectrum near a period doubling bifurcation. The "bump" structure just before the actual bifurcation is the noisy precursor.[47]

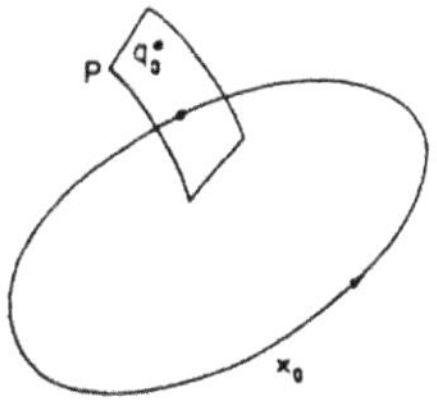

Figure 39. Stable period-one orbit. From Ref. 48.

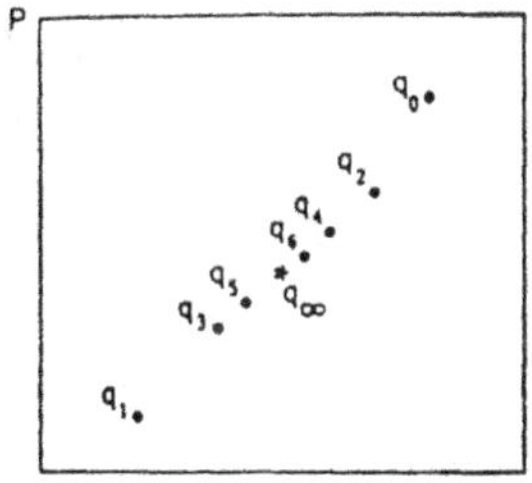

Figure 40. Successive intersections with Poincaré plane. From Ref. 48.

Another type of noisy precursor effect is the virtual Hopf phenomenon.[48] In this case, bump structure appears at two new frequencies, at $f_0 - f'$ and f'. As the control parameter is varied, f' increases, until the two bumps coalesce into a bump at $f_0/2$, much like the bump of a period doubling precursor. This is followed by a period doubling bifurcation.

Noisy precursor effects have been observed experimentally in solid state systems. For example, Fig. 42 shows the noisy precursor effect in a p-n junction system, near a period doubling bifurcation.[49] Figure 43 shows the virtual Hopf phenomenon for the CDW system $NbSe_3$.[50] The control parameter here is the dc bias voltage.

6.2 Mode Locking and Chaos

In Figs. 12 and 21 it was demonstrated that chaos can exist on mode locked steps. An intriging question is whether or not the mode locked region is nore or less likely to be sensitive to external noise. From the very terminology of a "locked" state, one might assume a resistance to external noise influence in the mode locked region.

We first examine the experimental situation for a mode locked CDW system. Figure 44a shows the differential resistance dV/dI and associated broad band noise for a dc driven CDW.[51] When the CDW

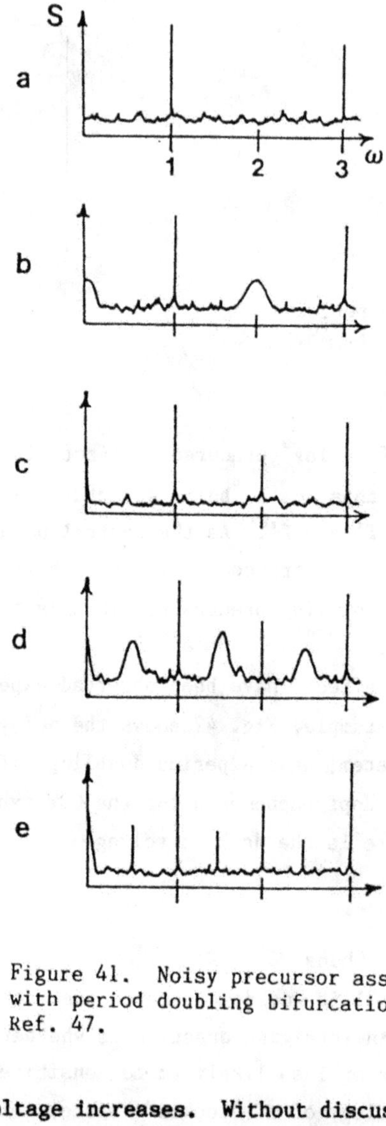

Figure 41. Noisy precursor associated with period doubling bifurcation. From Ref. 47.

moves, the noise voltage increases. Without discussing any microscopic reasons for such noise generation in CDW systems, we ask what happens to the noise amplitude when the CDW velocity is mode-locked. As shown

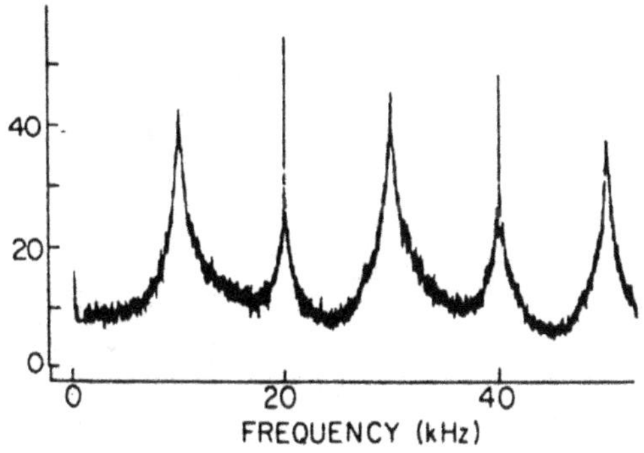

Figure 42. Noisy precursor in p-n junction.
From Ref. 49.

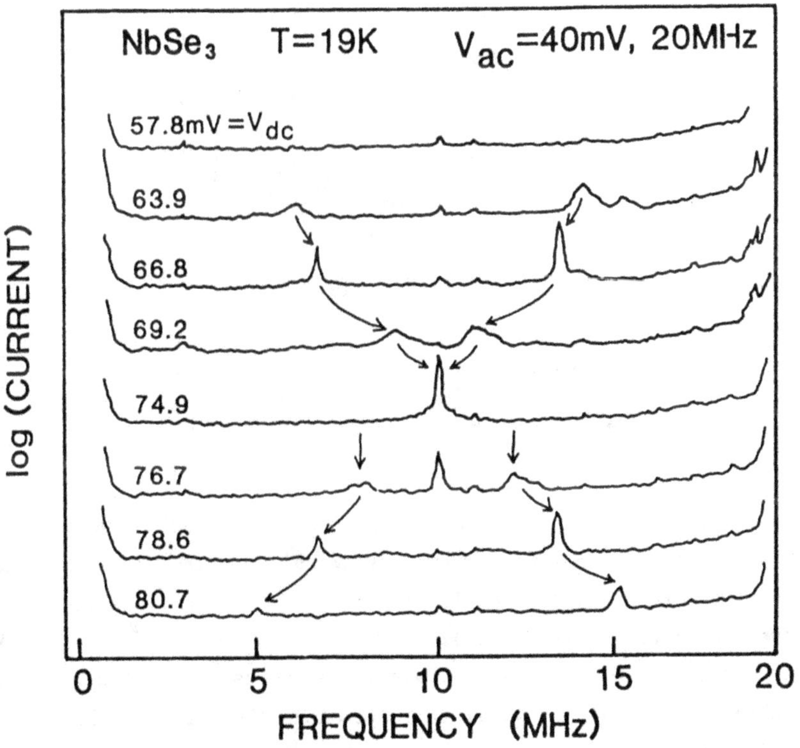

Figure 43. Possible virtual Hopf phenomenon in the CDW system $NbSe_3$.
From Ref. 50.

in Fig. 44b, during mode-lock the noise vanishes identically! In a sense, the locking has frozen out the degrees of freedom generating the conduction noise.

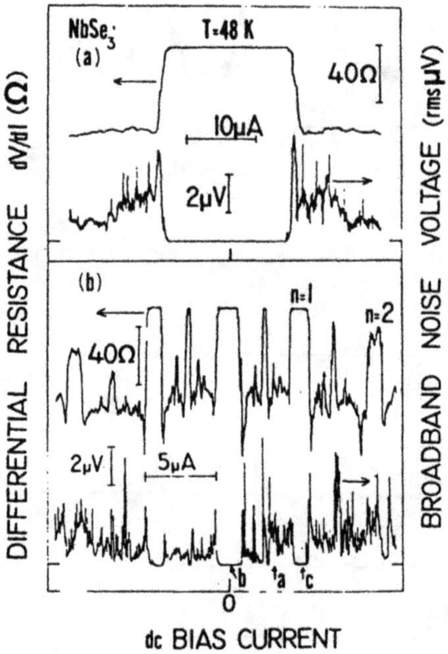

Figure 44. Mode locking and broad band noise in $NbSe_3$. a) no external ac drive; b) with external ac drive.

Wiesenfeld and Satija [52)] have suggested that these observations in the CDW system represent a generic dynamical problem in nonlinear mode locking. As a simple example they consider a Poincaré section on the invariant circle (circle map). Figure 45a shows a 3 to 1 locking; Fig. 45b shows an unlocked case. Consider dynamics <u>along</u> the circle. The locked case may viewed as dissipative along the circle, while the unlocked case is non dissipative along the circle. If the locked system is perturbed, the system is damped and relaxes back to the 3

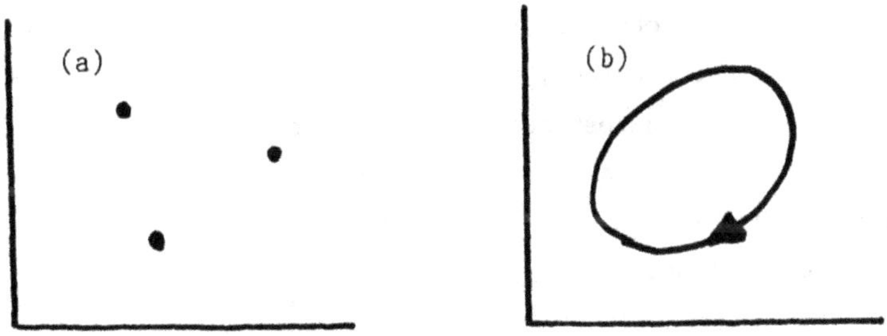

Figure 45. Circle map. a) 3 to 1 locking, fixed points shown. b) no locking. Adapted from Ref. 52.

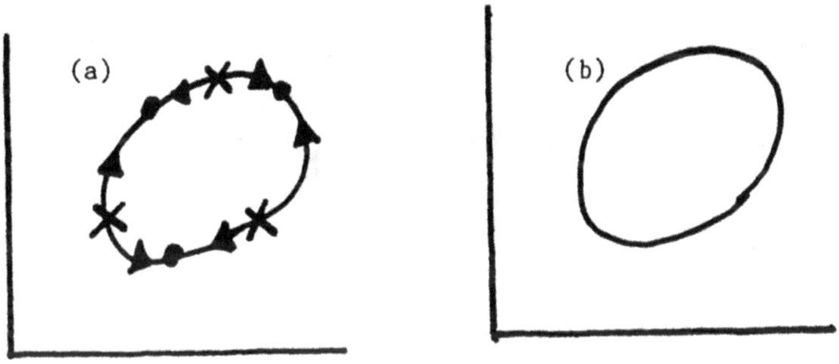

Figure 46. Dynamics along the circle map. a) locked regime: the perturbed system relaxes away from unstable fixed points to stable fixed points. b) no fixed points; system undergoes random walk about circle. Adapted from Ref. 52.

stable (and observable) fixed points (note that there exist also 3 unstable and unobservable fixed points; the system relaxes away from these). For the unlocked case, the 6 fixed points collide and annihilate in stable-unstable pairs. There is no relaxation of the perturbation, because no fixed points remain. This behavior is illustrated in Figs. 45a and 45b. In the unlocked case the noise produces a random walk away from the noise free orbit, and the perturbations accumulate with time. In contrast, in the locked case the noise produces perturbations which are damped out, leading to less sensitivity to external noise.

The CDW observations appear consistent with such an interpretation. In fact, the CDW system may be an ideally suited system in which to test more precise predictions of the theory of Weisenfeld and Satija, such as the response power spectra, etc. Mode locking and noise in general should provide a very interesting line of future chaos research.

ACKNOWLEDGEMENTS

It is a pleasure to acknowledge stimulating discussions and interactions with M.S. Sherwin, P. Bak, C. Jeffries, and M. Jensen during the course of this work. This research was supported by NSF grant DMR 84-00041, and the Alfred P. Sloan Foundation.

REFERENCES

[1] Eckman, J.P., Rev. Mod. Phys. 53, 643 (1981).

[2] Ott, E., Rev. Mod. Phys. 53, 655 (1981).

[3] Farmer, J.D., Ott, E., and Yorke, J.A., Physica 7D, 153.

[4] Swinney, H.L., Physica 7D, 3 (1983).

[5] Packard, H.H., Crutchfield, J.P., Farmer, J.D., and Shaw, R.S., Phys. Rev. Lett. 45, 712 (1980).

[6] Grassberger, P., and Procaccia, I., Phys. Rev. Lett. 50, 346 (1983).

[7] Huberman, B.A., and Crutchfield, J.P., Phys. Rev. Lett. 43, 1743 (1979).

[8] For a review, see A. Barone and G. Paterno, Physics and Applications of the Josephson Effect, (Wiley - Interscience, New York, (1982).

[9] Kautz, R.L., J. Appl. Phys. 52, 3528 and 6241 (1981).

[10] D'Humières, D., Beasley, M.R., Huberman, B.A., and Libchaber, A., Phys. Rev. A26, 3483 (1982).

[11] MacDonald, A.H., and Plischke, M., Phys. Rev. B27, 201 (1983)

[12] Octavio, M., Phys. Rev. B29, 1231 (1984).

[13] Goldhirsch, I., Imry, Y., Wasserman, G., and Ben-Jacob, E., Phys. Rev. B29, 1218 (1984).

[14] Huberman, B.A., Crutchfield, J.P., and Packard, N.H., Appl. Phys. Lett. 37, 750 (1980).

[15] Octavio, M., and Nasser, C.R., Phys. Rev.R30, 1586 (1984).

[16] For a review, see G. Gruner and A. Zettl, Phys. Rep. 119, 117 (1985).

[17] Gruner, G., Zawadowski, and Chaikin, P.M., Phys. Rev. Lett. 46, 511 (1981).

[18] Sherwin, M., Hall, R., and Zettl, A., Phys. Rev. Lett. 53, 1387 (1984).

[19] Shenker, S.J., Physica 5D, 405 (1982).

[20] Jensen, M.H., Bak, P., and Bohr, T., Phys. Rev. Lett. 50, 1637 (1983).

[21] Jensen, M.H., Bak, P., and Bohr, T., Phys. Rev. A30, 1960 (1984).

[22] Bohr, T., Bak, P., and Jensen, M.H., Phys. Rev. A 30, 1970 (1984).

[23] See, for example, B.B. Mandelbrot, The Fractal Geometry of Nature, (Freeman, New York, 1983).

[24] Jensen, M., et al., Phys. Rev. Lett. 55, 2798 (1985).

[25] Gwinn, E., and Westervelt, R., Phys. Rev. Lett. 57, 1000 (1986).

[26] Alstrom, P., Levinsen, M.T., and Jensen, M.H. Phys. Lett. 103A, 171 (1984).

[27] Miracky, R.F., Clarke, J., and Koch, R., Phys. Rev. Lett. 50, 586 (1983).

[28] Hall, R.P., and Zettl, A., Phys. Rev. B30, 2279 (1984).

[29] Brown, S.E., Mozurkewich, G., and Grüner, G., Phys. Rev. Lett. 52, 2277 (1984).

[30] Zettl, A., Physica 23D, 155 (1986).
[31] Coppersmith, S., and Littlewood, P., Phys.Rev.Lett. 57,1927(1986).
[32] Matsukawa, H., and Takayama, H., Syn. 19, 7 (1987).
[33] Hall, R.P., Hundley, M.F., and Zettl, A., Phys. Rev. Lett. 56, 2399 (1986).
[34] Feigenbaum, M.J., J. Stat. Phys. 19, 25 (1975); M.J. Feigenbaum, Physica 7D, 16 (1983).
[35] Nauenberg, M., and Rudnick, J., Phys. Rev. B24, 493 (1981).
[36] Crutchfield, J., Nauenberg, M., and Rudnick, J., Phys. Rev. Lett. 46, 933 (1981).
[37] Metropolis, N., Stein, M.L., and Stein, P.R., J. Comb. Theory Ser. A15, 25 (1973); see also Ref. 4.
[38] Testa, J., Pérez, J., and Jeffries, C., Phys. Rev. Lett. 48, 714 (1982).
[39] Buskirk, R., and Jeffries, C., Phys. Rev. A31, 3332 (1985).
[40] Grebogi, C., Ott, E., and Yorke, J.A., Phys. Rev. Lett. 48, 1507 (1982).
[41] Teitsworth, S.W., Westervelt, R.M., and Haller, E.E., Phys. Rev. Lett. 51, 825 (1983).
[42] Teitsworth, S.W., and Westervelt, R.M., Physica 23d, 187 (1986)
[43] Teitsworth, S.W., Westervelt, R.M., Phys. Rev. Lett. 53, 2587 (1984).
[44] Hoh, F.C., Lehnert, B., Phys. Rev. Lett. 7, 75 (1961).
[45] Held, G.A., Jeffries, C., and Haller, E.E., Phys. Rev. Lett. 52, 1037 (1984).
[46] Held, G., University of California Ph.D. Thesis (1985).
[47] Wiesenfeld, K., J. Stat. Phys. 38, 1071 (1985).
[48] Wiesenfeld, K., Phys. Rev. A32, 1744 (1985).
[49] Jeffries, C., and Wiesenfeld, K., Phys. Rev. A31, 1077 (1985).
[50] Zettl, A., Sherwin, M.S., and Hall, R.P., Physica 143B, 69 (1986).
[51] Sherwin, M.S., and Zettl, A., Phys. Rev. B32, 5536 (1985).
[52] Wiesenfeld, K., and Satija, I., Phys. Rev. B36, 2483 (1987).

NONLINEAR DYNAMICS AND CHAOS IN OPTICAL SYSTEMS

N.B. Abraham
Department of Physics
Bryn Mawr College
Bryn Mawr, PA 19010, U.S.A.

Two paradigmatic optical systems are considered as instructive examples from which we can learn about nonlinear dyanamics and chaos. In the first example, we will study a classic example of a nonlinear optical device with feedback which results in multiple steady state solutions for the same operating conditions. This system is the most frequently presented as an example for which the dynamics is well understood. Period doublings and chaotic behavior have been observed as predicted, but some subtle differences between experimental results and the theoretical predictions have forced revisions in the simple models that were initially put forward to explain the behavior. The second example is the case of a laser operating on a single mode. Here again the predictions are relatively simple and straightforward and we learn how the nonlinear field-atom interaction can lead to periodic and aperiodic modulation of the output. Although many lasers show periodic and chaotic modulation, experimental realizations of this model have been difficult to achieve. There is increasing evidence that there soon may be excellent working systems to compare with the predictions of the models.

CONTENTS

1. INTRODUCTION

It is no longer surprising that the output of many optical systems is not easily stabilized; it is natural to expect intensity fluctuations, if only because of the intrinsic spontaneous emission. However, in many laboratory applications, it is routinely possible to achieve excellent amplitude and frequency stability. It is with this in mind that we take as a starting point the ideal case of an essentially noise-free optical system, the parameters of which are held constant. The natural expectation is that the system will reach a steady state in which the output is also constant. While such solutions almost always exist, they are often not stable with respect to infinitesimal perturbations which drive the system toward a time-dependent form of behavior. This behavior is then called "self-pulsing" or spontaneous pulsation and is the manifestation of what we more generally call nonlinear dynamics.

The pulsations can be regular or irregular, but as we have assumed that there is no noise in the system, the behavior is always deterministic. Irregular deterministic behavior may appear to be a contradiction in terms, but it is not, and the field of study of nonlinear dynamics and chaos has grown up to deal with these kinds of problems. A variety of excellent reviews and new books have recently appeared and can be used to give the reader an overview of the problem.[1-13)]

In these notes, we will move directly to a discussion of the behavior of several optical systems so that the special kinds of time-dependent output can be pictured as resulting from "real" situations.

2. BACKGROUND ON OPTICAL INSTABILITIES

The study of the dynamics of nonlinear optical systems is often referred to as the study of "optical instabilities". As "optical" in such usage is no longer limited to the visible portions of the spectrum, the reader is cautioned that the term will apply to dynamical behavior in many different forms of the interaction of electromagnetic radiation with matter. Interesting systems which have been studied include those with wavelengths corresponding to X-rays and those corresponding to radio waves, a span of more than seven orders of magnitude in frequency and wavelength.

Although the history of the study of optical instabilities for lasers and masers began in the late 1950s[14], the field did not become fully developed until the last five years. Two principal foci are optical bistability and lasers. Optical bistability refers to systems which have more than one (literally two) steady state solutions for a given set of operating conditions. Of course, from our point of view, it is the dynamics of a bistable system which will be of special interest. This area began to open up in the late 1970s[15,16] and since 1980 has been an area of very active investigation. The laser instability work remained a subject for theorists until the late 1970s and has since grown to be a very active field as well. There have been many meetings recently on this subject and several proceedings volumes and other compendia of research articles give reviews of the history and reports of the present activities in the field.[17-23]

3. A BISTABLE OPTICAL DEVICE

The term optical bistablity has been applied to optical devices which are naturally absorbing for small incident intensity, but the absorption becomes nonlinear for stronger input. Thus if there is to be any output, there must always be an incident field. There have been two classic designs for studies of optical bistability; examples are shown in Fig. 1.

a)

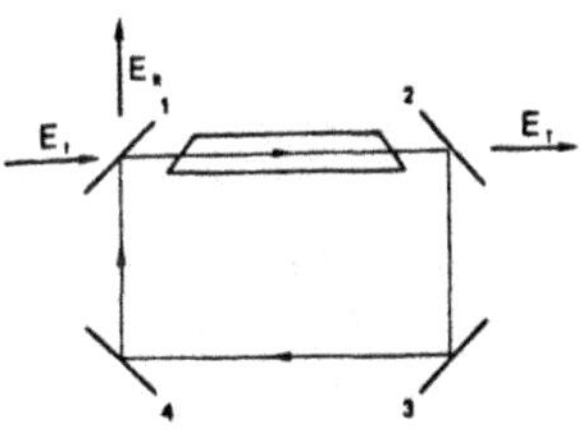

b)

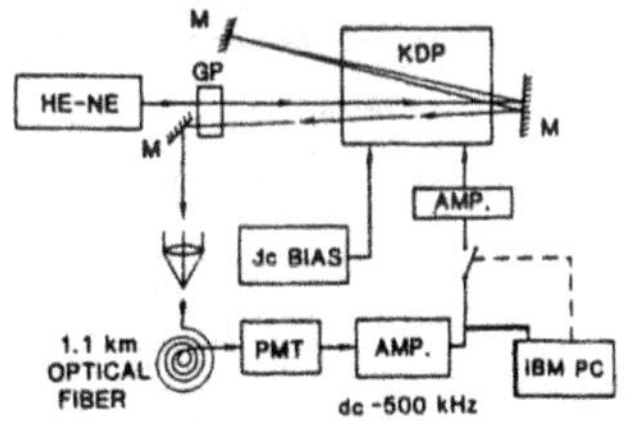

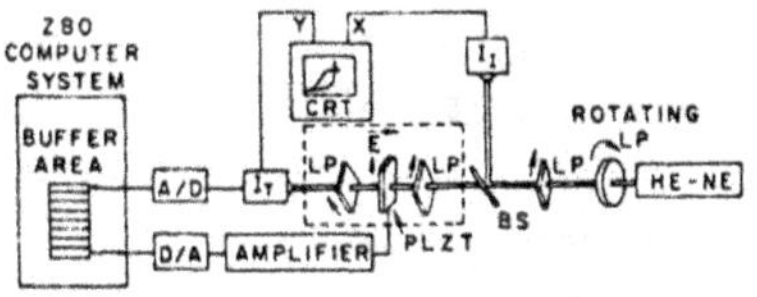

Fig. 1: a) an schematic (after Ref. 27c) of an optical ring cavity with a nonlinear absorbing medium; b) two experimental setups of hybrid nonlinear optical systems with delayed feedback using a digital delay line[26a] and a fiber optic delay line[26c].

The first may be thought of as the more directly physical for dealing with an "all optical" system, the second permits more general kinds of (not necessarily optical) feedback and is rather more easily described and built experimentally. Another important difference is that the ring cavity is sensitive to the phase of the field that is cycled back by the cavity while the hybrid device is sensitive only to the intensity. The first can be modelled specifically to describe a set of two-level atoms in a homogeneously broadened medium. The nonlinear partial differential equations for the field and atomic variables are given by Eqs.(3.1), which can be simplified by assuming a plane wave in the cavity and a uniform field along the direction of propagation. The equations to describe the ring cavity system as given by Lugiato[24] are

$$\frac{\partial E}{\partial t} + c\frac{\partial E}{\partial z} = gP ,$$

$$\frac{\partial P}{\partial t} = \frac{\mu}{h} ED - [\gamma_{\perp} + i(\omega_A - \omega_0)] P , \tag{3.1}$$

$$\frac{\partial D}{\partial t} = -\frac{\mu}{2\hbar}(EP^* + E^*P) - \gamma_{\parallel}[D - (N/2)] ,$$

where E is the field amplitude, P is the atomic polarization, and D is the population difference, all of slowly varying amplitudes with respect to the carrier wave at ω_0, and ω_A is the resonant frequency of the atoms. The closed cavity with partially transmitting mirrors gives the boundary condition

$$E(0,t) = T^{1/2} E_I + R \exp\{-i(\omega_C - \omega_0)\ell/c\} E(L,t-\Delta t), \tag{3.2}$$

where E_I is the amplitude of the incident field, L is the length of the medium and ℓ is the length of the resonator path minus the length of the medium, T is the transmitivity of the input mirror, R (= 1-T) is the reflectivity of the input and output mirrors, and ω_C is one of the resonant frequencies of the empty cavity. (For convenience and to reduce the amplitude fluctuations as much as possible we select the cavity resonance closest to the input frequency ω_0.)

The equation to describe an electro-optical system with delay as given by Ikeda,[16] Gibbs,[25] Hopf et al.,[26] Gao and Narducci,[27] or Vallée and Delisle[29] is

$$\gamma^{-1} \dot{\beta}(t) = -\beta(t) + A^2[1 + 2B\cos\{\beta(t-t_R) - \beta_0\}], \qquad (3.3)$$

where $\beta(t)$ is a normalized intensity, t_R is the delay time in the feedback loop, β_0 is a bias voltage applied to the electro-optical modulator, B and A are various amplification and normalization factors, and γ is the decay rate of the intensity in the loop. $T \equiv \gamma t_R$ is the important parameter which measures the length of the delay in units of the medium response time.

Under very limited conditions, it is possible to show that the ring cavity problem can be reduced to the system-with-delay problem. That this is not trivial is revealed by the amount of controversy that arose over this point. It was clarified by discussions of Narducci and Lugiato[30] and Gibbs.[25]

We wish to focus on the system with delay, because it is a schematic design which can easily be achieved in the laboratory and which has been studied with nonlinear media of electro-optic crystals,[26,31] acousto-optic crystals,[29,32] and liquid crystals[33] among others, and with delay lines generated by digital computers,[26,31,33] optical fibers,[26c,39] and simpler optical paths. The model with delay is also of interest because it was in studying it that Ikeda and coworkers first found dynamical chaos in a model for a bistable optical system. His reports and predictions sparked much of the vigorous experimental research on systems of this type that has followed since.

It is worth noting, however, that the optical ring cavity filled with a two-level atomic medium is a theoretical paradigm that has also been realized experimentally in the beautiful experimental work of Kimble and coworkers[34] for single-mode effects and by Macke and coworkers[35] for multimode dynamics. The fits between their data and various theoretical analyses of the model equations have been very exciting. Thus both models and their corresponding experimental systems have been subjects of intense study over the last five to seven years.

Equation (3.3) for the system with delay is called a "delay-differential equation". We see both the time derivative term and the coupling to the value of the intensity at an earlier time. One of the simplifications that has been studied extensively results from the assumption that the time derivative term could be made negligibly small by a suitable selection of the response-time of the medium. It is argued that if γ^{-1} is sufficiently small, the medium is able to adjust instantaneously to the field and so the time derivative of β is negligible. The validity of this assumption will be examined later. The result of setting the derivative equal to zero is to formulate a delay equation (sometimes also called a map) that relates the value of the intensity at time t to the value exactly one delay time later:

$$\beta(t) \;=\; A^2\,[1 + 2B\cos\{\beta(t-t_R) - \beta_0\}]. \qquad (3.3)$$

In this picture, the intensity can be represented by a sequence of values equally spaced in time by the value of the roundtrip time t_R; each value is predicted exactly by the immediately preceeding value, but only a single value of the signal during each roundtrip time is thereby predicted.

In either case, the first step in understanding the system lies in examining steady state solutions. One assumes a constant solution, setting the time derivatives equal to zero (or assuming that ß is time-independent), and solves for the allowed values of the output intensity in terms of the input intensity. For suitable values of the parameters, the output curve becomes triple-valued or multivalued as shown in Fig. 2.

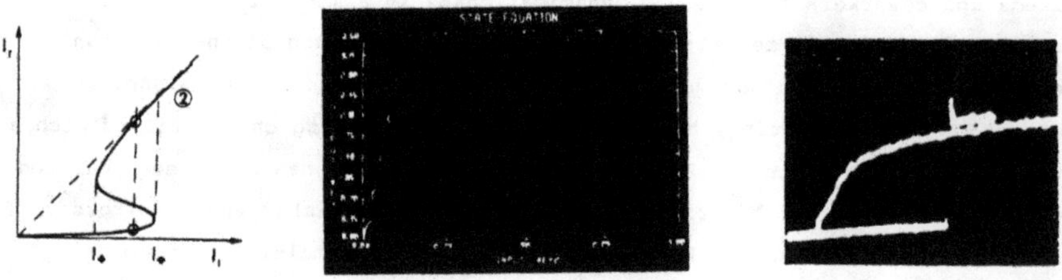

a) b) c)

Fig. 2. Intensity output vs. intensity input curves for the ring model (a)[27c], for the hybrid circuit model (b)[27a], and from the first hybrid experiment (c)[26a] showing only the stable portions of the curve.

As the portion of negative slope can be shown to be unstable with respect to perturbations on simple physical grounds, we are left with two (or more) possible steady states; hence the general terminology of optical bistability. When both of these states are stable, the region of bistability gives rise to optical switching and a kind of optical logic wherein we can find two states (high intensity output and low intensity output) for the same input intensity. The selection of one of the two states depends on the past history of the device.

For the map, we can also find certain periodic solutions by a clever trick: n-periodic solutions are the fixed points of the map composed on itself n times.

The stability of a solution can be tested analytically by perturbing it and determining the response of the new solution. If the perturbation decays away, then the solution is stable. If the perturbation grows, then the solution is unstable. If all steady state solutions are unstable, then there is surely at least one time-dependent solution. However, there may often be more than one stable solution, a condition called "multistability" and the selection of a solution will then depend on the initial conditions.

When the stability of a continuously varying system is being examined, often the problem can be linearized about the steady state solution and an exponential growth factor can be assumed. Solving the stability problem then reduces to solving a secular equation for the eigenvalues of the system. If the real parts of all of the eigenvalues are negative, then the solution is stable. If one of the eigenvalues has a positive real part, then the solution is unstable. Complex eigenvalues suggest that the growth or decay of the perturbations will be oscillatory.

For the corresponding map, the solution can only be tested after one iteration. One speaks then of Floquet exponents which are the logarithm of the deviation from the steady state solution after one iteration. If the norm of all of the exponents is less than one then the steady state is a stable fixed point. If the norm of any one of the exponents exceeds one then the steady state is unstable.

The map appears to work well as shown in Fig. 3a where the predicted sequences of values are compared with the experimental observations. The regions marked 1, 2, and 4 indicate the number of roundtrip times before the pattern repeats. A single value is taken each rought trip time from the experimental data. A constant signal is labelled 1, repeating every roundtrip time. A simply periodic solution is labelled 2 because it switches between two values and then repeats.

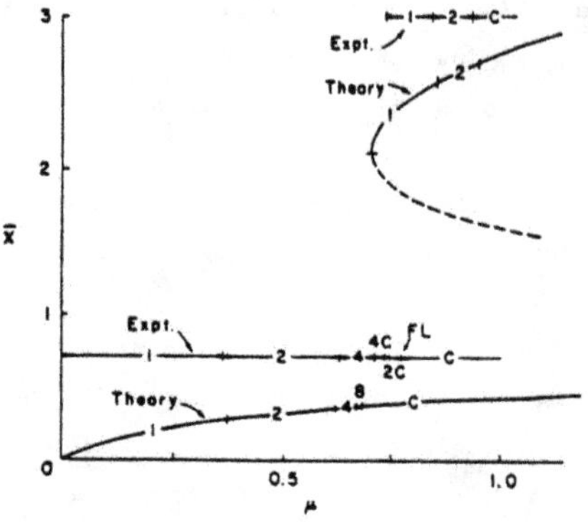

Fig. 3a. Comparison of experimentally observed patterns in the hybrid with the predictions from the map following Hopf et al.[26b)] X is our ß and μ is our A.

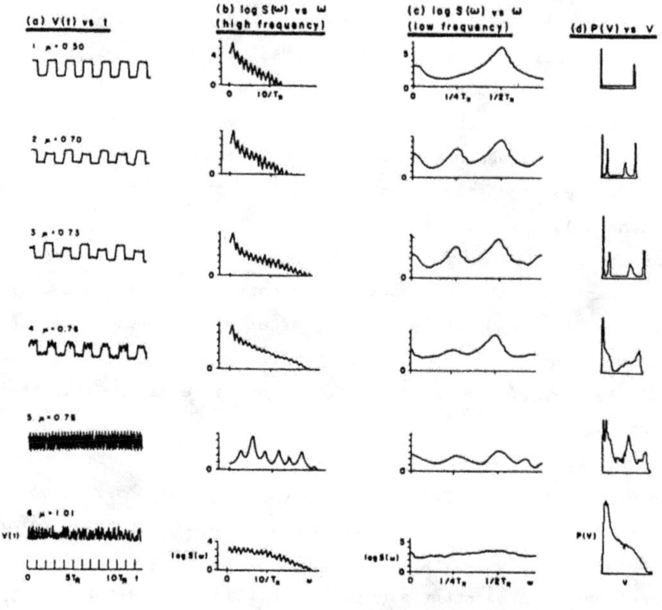

Fig. 3b. Various experimental output waveforms and characteristics following.[26b)]

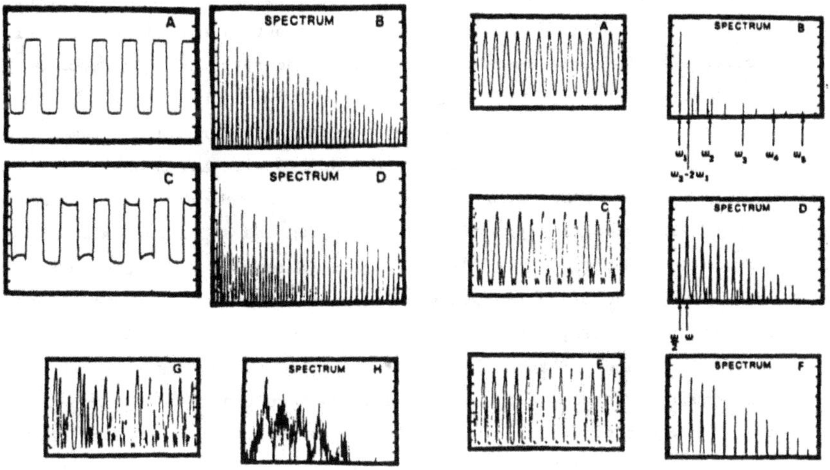

Fig. 3c. Numerical solutions of the delay-differential equations.[27c)]

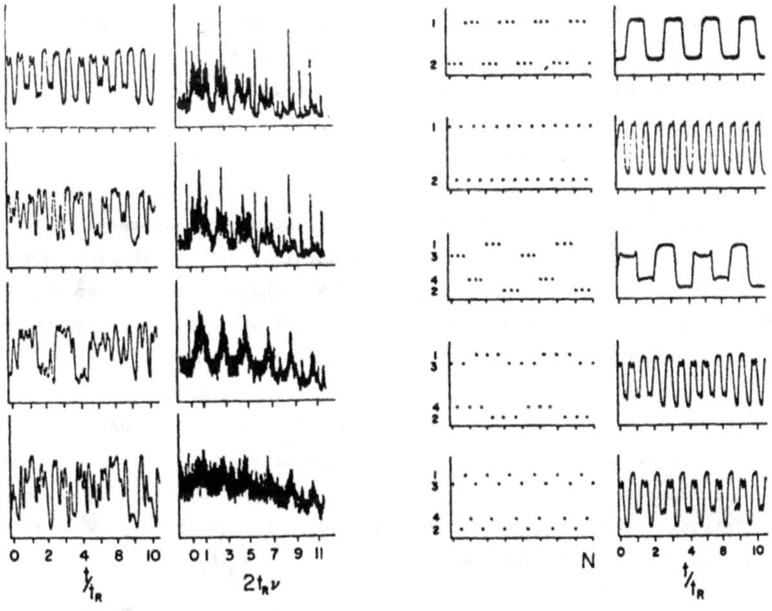

Fig. 4. Different forms of chaotic behavior and comparisons of higher order frequency lockings in a hybrid system with interspersed maps.

Sample time patterns, their associated spectra, and their probability distribution functions are shown in Fig. 3b for experimental measurements by Hopf et al.[26b] Solutions to the delay differential equations and their associated spectra are shown in Fig. 3c from the results of Narducci et al.[27c]

Patterns from Derstine et al.[26c] showing experimental results for different kinds of chaos, more complex periodic waveforms, and attempts to model them using three interspersed maps per roundtrip time are shown in Fig. 4.

We see that there is a variety of pulsation behavior that can be observed. The simplest type of oscillation repeats after two delay times. This pulsation is sometimes called "$2t_R$" because it repeats after two roundtrip times. Period doubling is observed in which the pulsation pattern becomes more complicated, repeating after four or eight or sixteen roundtrips. The regions of parameter space for successive period-doubled behavior shrink until the period-doubling converges to an irregular pattern that is called chaotic behavior. The convergence is of the sort that has been studied by May[36] and Feigenbaum[37] and which has been shown to be universal in that the ratio of certain parameter differences which span successive types of periodic waveforms approaches a constant and the heights of successive new peaks in the power spectra approach a constant ratio as well.[37] These ratios have been shown to be the same for a large family of dissipative systems.

In the chaotic region, the system, though evolving in a nonperiodic fashion, is nevertheless limited to a relatively small subset of all possible sequences of values. These can be represented by trajectories in the phase space of the system. The sequences of values followed asymptotically are called an attractor because arbitrary initial conditions converge toward a common set in the long term. However, arbitrarily close initial points diverge exponentially from each other, a feature that leads us to call the attractor a "strange attractor". Hence, although exact prediction is possible, the sensitive dependence on initial conditions makes impossible long-term forecasting based on only approximate information (no matter how small the initial uncertainty might be). In the chaotic region the power spectra have a continuous portion, in marked contrast to the narrow peaks associated with periodic solutions. In this particular system, there are also other periodic signals that can be observed within the chaotic region; these are cases of locking of the switching frequency to odd multiples of $1/2t_R$.

Two useful ways to present the behavior of the system are shown below in Fig. 5 and Fig 6. In Fig. 5, the values of the solution (the values from the ma p or selected values at fixed delay time for the continuous data trace) are plotted on the vertical axis and the value of one of the parameters of the system is plotted on the horizontal axis. For each different value of the control parameter, the nature of the solution is revealed by the number of points on the corresponding plot. A finite number of points indicates a periodic solution, a blurr suggests the irregular, chaotic solution. This type of presentation is called a "bifurcation diagram".

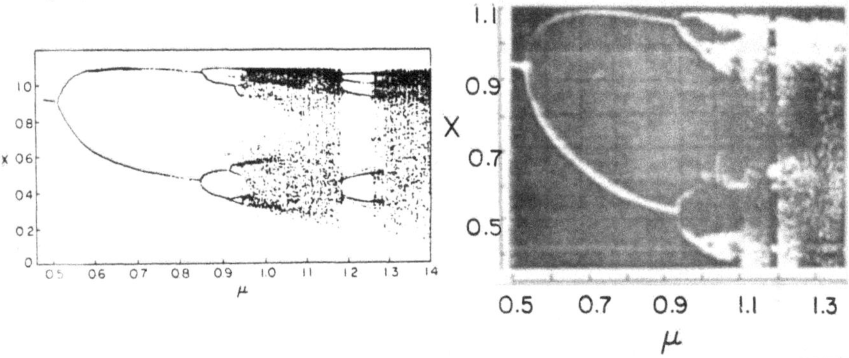

Fig. 5. Bifurcation diagrams from theoretical and experimental work by Vallée and coworkers.[29a)]

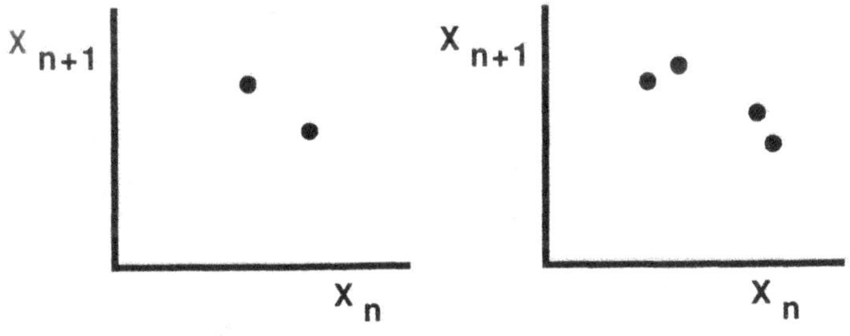

Fig. 6. Schematic next value plots for a periodic ($2t_R$) solution and a period-doubled ($4t_R$) solution.

In Fig. 6, a "next-value" map is plotted, the horizontal coordinate is a particular value in the sequence and the vertical coordinate is the corresponding next value in the sequence. When the solution is periodic, there is a finite number of points in this next-value map. When the solution is chaotic, the points are still restricted to a subregion and the points seem to remain distinct. An investigation of the resulting pattern reveals that the points form a set of fractional dimensionality, called a fractal. This characteristic feature of chaotic solutions is one of the easiest to measure. Beginning from a sequence of values, it is now quite common to compute the dimensionality of the attracting set in order to confirm chaotic behavior.

The period-doubling sequences that are recorded in experiments are often limited, instead of showing all of the infinite number of the successive higher periodicities before chaos is observed. This is often attributed to the narrowness of the regions in parameter space in which these would be observed and the fact that any small amount of noise would blur out the solutions by jittering the control parameters. This has been explored theoretically and experimentally by Crutchfield and Hubermann[38)] and by Derstine, Gibbs, and Hopf.[39)] The experimental results are substantially in accord with the theory.

Despite very good agreement in many cases between the experimental results and the predictions from the difference equations, the approximations made to reach the difference equation have been demonstrated to be invalid. Intuitively, when one sees the experimental result, one is inclined to question the neglect of the differential term during the sharply rising and falling edges of the nearly square-wave solutions. In fact, if one turns to the limiting case that might best justify the neglect of the differential term, we see in the solutions of the delay-differential equation that square-wave rather than smoothly varying solutions are more common. It turns out that the map is only exact at predicting the first transition from constant to periodic behavior. The map is unable to represent the complexity during one roundtrip period as it can supply only a single value: it can represent neither the sharp rises and falls nor the detailed structure.

Several examples of solutions from the delay-differential equations for values of T near 1.0 are shown from the work of Gao and Narducci[27a)] in Fig. 7. These have recently been observed in experiments on an electro-optical hybrid device.[31)] Among the most interesting results of the general study of the delay-differential equations is the truncation of the period-doubling sequences by naturally occuring dynamical mechanisms rather than by noise. This was first noted by Gao and Narducci[27)] in 1983 and elaborated upon by Mandel and

Erneux.[28] More recently, Le Berre, Ressayre, Tallet and Gibbs[40] also noticed this effect for a two-dimensional case (delay differential equations for a ring cavity where the system is sensitive to the phase of the electric field) and have shown that the detailed structure of the attractors from the two models can appear to be similar but actually may differ widely in several properties. Some selected examples are shown in Fig. 8. The attractors of the delay differential systems have much higher dimensions than those limited values possible for the map.

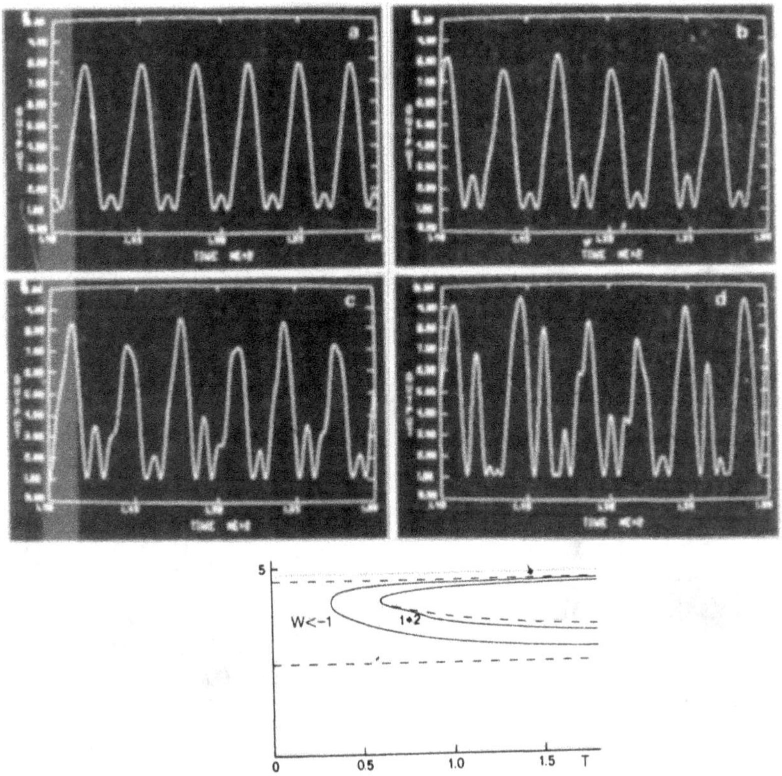

Fig. 7. Results from Gao and Narducci[27a] for solutions of the delay differential equations with T of order 1 (small delay times). Also a phase diagram showing regions of transitions directly from periodic to chaotic waveforms without period-doubling where the period-doubling boundary labeled 1->2 intersects the dotted line boundary of chaos.

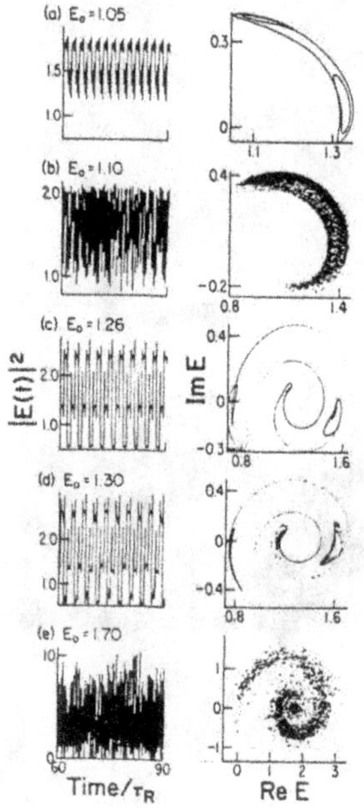

Numerical integration of Eqs. (1), (2), and (3) with $R = 0.95$, $\alpha l = 4$, $\Delta = 3\pi$, and $\gamma\tau_R = 10$. No period-doubled wave form was found between (a) and (b) for 10^{-3} steps in E_0 between 1.065 and 1.08.

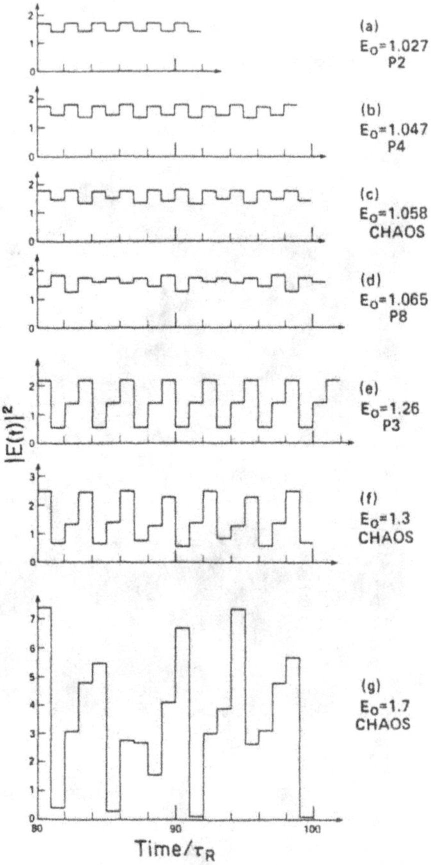

Bifurcation sequence for the 2D mapping with same parameters as Fig. 1. A period-doubling route to chaos occurs between (b) and (c). Periodic windows with the chaotic domain also occur; see (d) and (e).

Fig. 8a. Results from Le Berre <u>et al.</u>,[40a] comparing waveforms from the 2D map with those from the delay differential equation for the same parameter values.

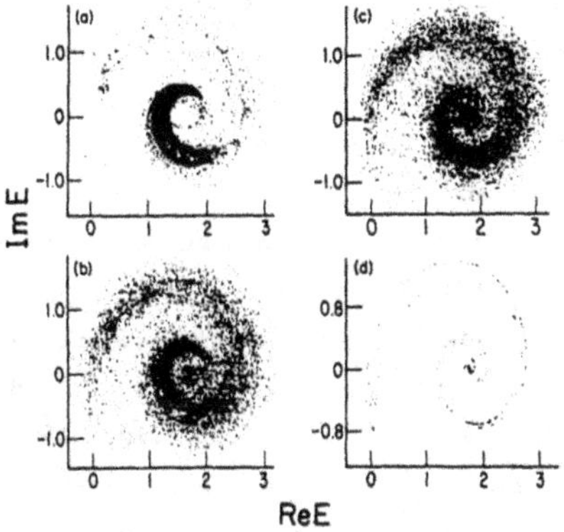

Fig. 8b. Results from Le Berre and coworkers[40a] on chaotic attractors for their delay differential equation (a-c) with T = 1,2,5, respectively, and for the corresponding 2D map approximation (d) which is often presumed incorrectly to correspond to T = ∞.

At this point it is fair to say that use of delay equations (or "maps") is no longer considered the proper procedure. However, because of their long history, one can be sure that study of these approximations will continue, (see for example Ref. 41), and the results should be viewed with considerable caution.

Details of the delay-differential equations show another interesting feature. The square wave structure of the delay equation solutions can be retained under the condition that many eigenvalues of the problem go unstable simultaneously and that their imaginary parts are exactly odd multiples of the fundamental. This leads to the phase-locking of the "modes" and the formation of the square-wave solutions. For smaller time delays, the imaginary parts of the eigenvalues of the stability analysis problem are not rationally related, hence quasiperiodic behavior occurs more easily. Lists of the eigenvalues with the largest real parts for several different operating conditions have been published by Gao and Narducci[27] and the imaginary parts have recently been observed as peaks in the power spectrum of the pulsing system. The differences between harmonics of the fundamental frequency and the imaginary parts of the other eigenvalues have been highlighted by Vallée and coworkers.[29]

These sorts of pulsations that involve the roundtrip time in the cavity or in the feedback loop, can be properly called multi-mode instabilities, in that the field can be decomposed into a set of fields resonant with the boundary conditions whose frequencies differ by the speed of propagation divided by the roundtrip distance.

What are properly called single mode instabilities are modulations of the amplitude of a single resonant mode. These do not occur for the delay differential equations. Such behavior would be possible in a ring cavity configuration such as designed in Fig. 1a and as described by nonlinear field-matter equations such as Eqs. (3.1). Typically these modulations arise from the various nonlinear couplings between the field and the medium and the periods are not locked to the cavity roundtrip time. In principle, the feedback-loop length can be varied to separate single-mode from multi-mode pulsations. In practice the separation is not always so easy. In consequence there is often confusion about this terminology in the literature. Single mode pulsations have been seen in optical bistability experiments in a ring resonator as reported in Ref. 34.

4. A SINGLE MODE LASER

The single mode laser, shown schematically below in Fig. 9, is conceptually similar to the absorber in the ring cavity. The only changes are to replace the absorber with a material from which net optical energy can be extracted by stimulated emission and to remove the external signal.

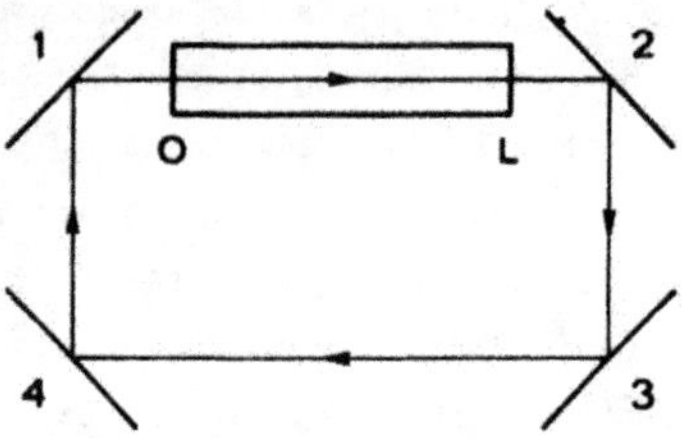

Fig. 9. Schematic of a unidirectional ring laser.

In the simplest approximation, the field in this cavity is assumed to have a uniform transverse profile (the plane-wave approximation) and to be uniform along the optical path within the cavity. (This is now known as the "uniform field approximation", a change from its previous appellation as the "mean field approximation" which was confusing because it has nothing to do with the mean field approximations of statistical mechanics.) If the field amplitude is slowly varying on the length scale of a wavelength and on the time scale of an optical period, then the Schrödinger equations for a two level medium and the Maxwell equations for the field can be reduced to a set of three coupled ordinary differential equations:

$$\frac{dE}{dt} = -\kappa[E + AP] \,,$$

$$\frac{dP}{dt} = -\gamma_\perp[ED + \{1 + i(\omega_A - \omega_C)/\gamma_\perp\}P] \,, \tag{4.1}$$

$$\frac{dD}{dt} = -\gamma_\parallel[(1/2)(E^*P + P^*E) + (D - 1)] \,.$$

A version of these equations was first presented by Oraevskiy and others[14,42] and a formal derivation and discussion can be found elsewhere.[23]

Solution of these equations begins, as before, with the identification of the steady state solutions. Of physical interest are the zero-field state and the nonzero field state. These are given, respectively, by:

Zero field:

$E = 0, \quad P = 0,$

$D = 1,$

Nonzero field:

$|E_{ss}|^2 = A - 1 - \Delta^2,$

$D = (1 + \Delta^2)/A,$

where $E = E_{ss}\exp[i(\omega_A - \omega_C)\kappa t/(\gamma_\perp + \kappa)]$ and $\Delta = (\omega_A - \omega_C)/(1 + \kappa/\gamma_\perp)$.

As we are often interested in the change in the solutions for the laser system as we change the level of excitation of the medium, it is often helpful to plot the solutions as functions of the threshold parameter A as is done in Fig. 10.

For sufficient excitation of the medium, both states exist. However, when the nonzero-intensity solution exists, the zero-intensity solution is unstable. This means that we can examine the stability and dynamics of the laser by studying the stability of the nonzero-intensity solution.

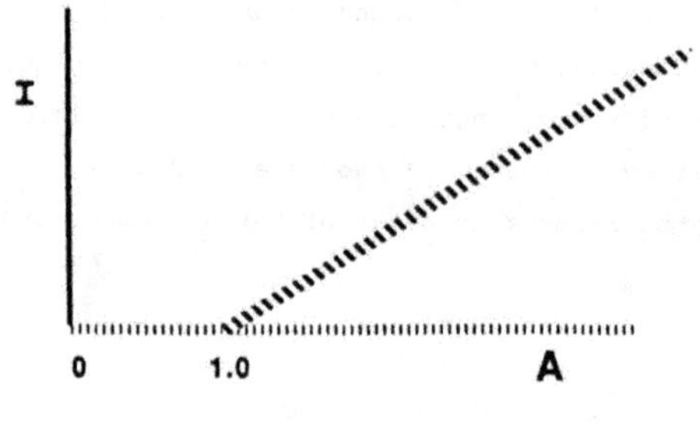

Fig. 10. Plot of laser intensity for steady states.

Perturbing the steady state solution, linearizing the equations with respect to the perturbations, and seeking exponential solutions leads to the secular equation[23)] for the growth rate λ:

$$\lambda\{\lambda^4 + (2\gamma_\perp+2\kappa+\gamma_\parallel)\lambda^3 + [2\gamma_\parallel(\gamma_\perp+k) + (\gamma_\perp+\kappa)^2 + (I\gamma_\perp-\gamma_\parallel)\gamma_\perp{}^2 + (\kappa-\gamma_\perp)^2\Delta^2]\lambda^2$$

$$\gamma_\parallel[(\gamma_\perp+\kappa)^2 + (3\kappa+\gamma_\perp)I\gamma_\perp + (\kappa-\gamma_\perp)^2\Delta^2]\lambda + 2\kappa\gamma_\parallel(\gamma_\perp+\kappa)I\gamma_\perp\} = 0 \quad .$$

Solutions of this fifth-order polynomial equation give the eigenvalues of the problem and the corresponding eigenvectors give the directions in the five-dimensional phase space for the characteristic evolution of the perturbed solution. The zero eigenvalue can be shown to correspond to the absolute phase of the laser field, which is not specified by the steady-state solution. The remaining four eigenvalues govern the behavior of the coupled field, polarization, and population inversion variables.

As the excitation level is increased, we find that two of the eigenvalues are complex conjugates with negative real parts which increase toward zero. If certain conditions are met, namely,

$\kappa > \gamma_\perp + \gamma_\parallel$ and

$A > 1 + \{(\gamma_\perp + \gamma_\parallel + k)(\gamma_\perp + \kappa)\}/(\kappa - \gamma_\perp - \gamma_\parallel)$, the real parts of these eigenvalues cross from negative to positive with increasing excitation, causing the constant intensity solution to become unstable with respect to pulsations. This kind of loss of stability (as in the optical bistability case) is called a Hopf bifurcation, where "bifurcation" refers to the branching to the solutions at this point to include a new type of behavior. At the most favorable conditions for an instability, the cavity decay rate should be about three times the sum of the field and polarization decay rates and the excitation must be at least 10 times above threshold.

To see more of the interesting types of behavior of this system, we must turn to the numerical solution of the equations. Various kinds of pulsations have been observed. Samples are presented in Fig. 11.

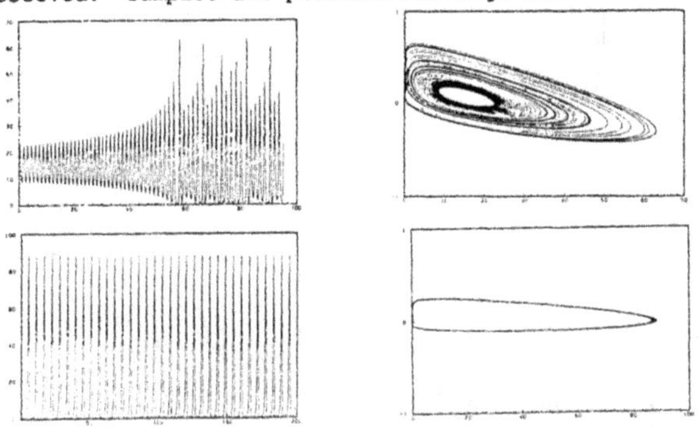

Fig. 11. Sample time dependent behavior of the laser intensity and plots of the waveform in the variable space of I vs. D.

To understand the nature of some of these pulsations, it is instructive to plot a phase-space projection. Then we can observe the solution with time as a parameter. We see in Fig. 12 that sometimes the solutions are symmetric with respect to a parity inversion of the electric field (the equations have this symmetry) and sometimes the solutions are asymmetric. The asymmetric solutions may result in intensity pulsations that appear to be period-doubled because successive pulses alternate in intensity. However, we should not properly call this a period doubling because in the fundamental variable space the trajectory is a single closed curve after one revolution.

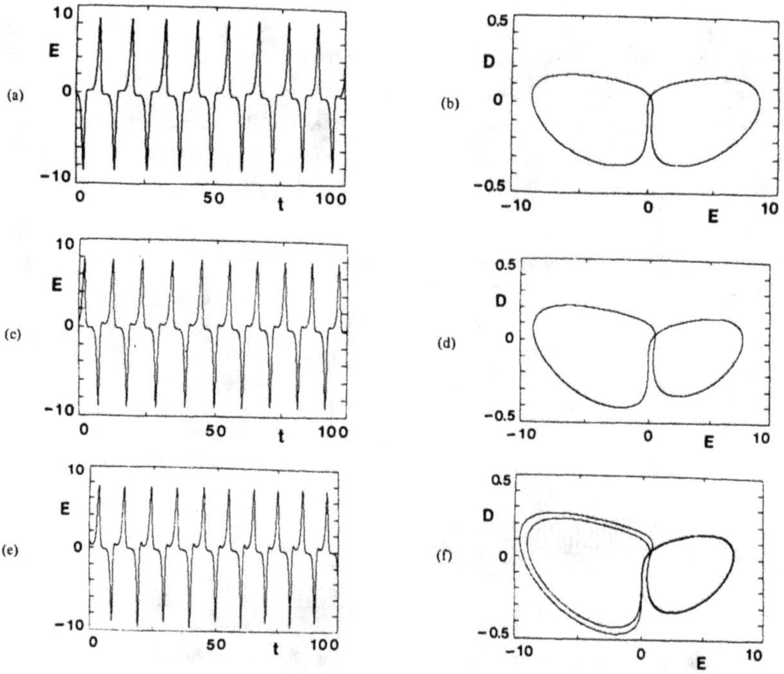

Fig. 12. Time-dependent electric fields and phase-space pictures showing symmetric, asymmetric, and period-doubled cases.[43)]

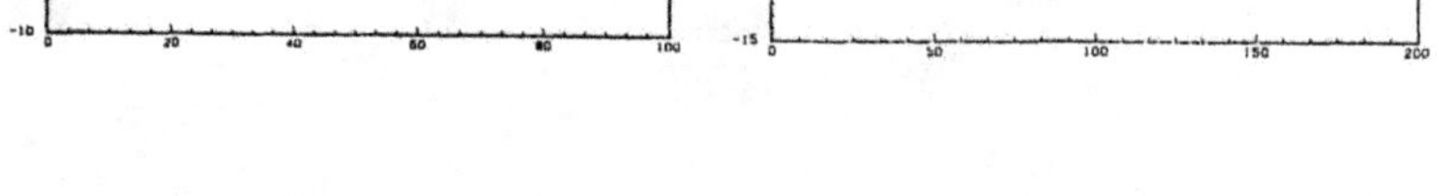

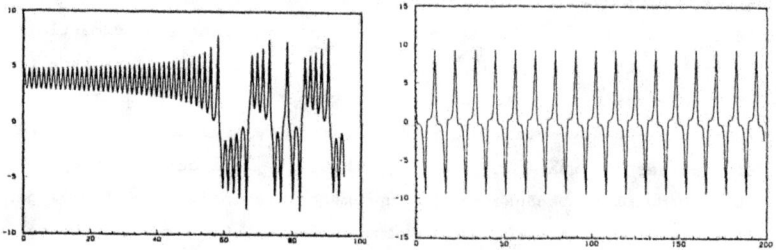

Fig. 13. Periodic and chaotic solutions.

Most often for this single-mode laser model, the solutions are chaotic rather than periodic if the laser is tuned to resonance. However, this does depend on the parameters, and periodic solutions have been found near the instability threshold in some cases.

With detuning of the cavity resonance from the atomic resonance frequency, periodic solutions are much more common. An example is shown below for a bifurcation diagram where the peak values of the intensity are plotted for different values of the cavity detuning.

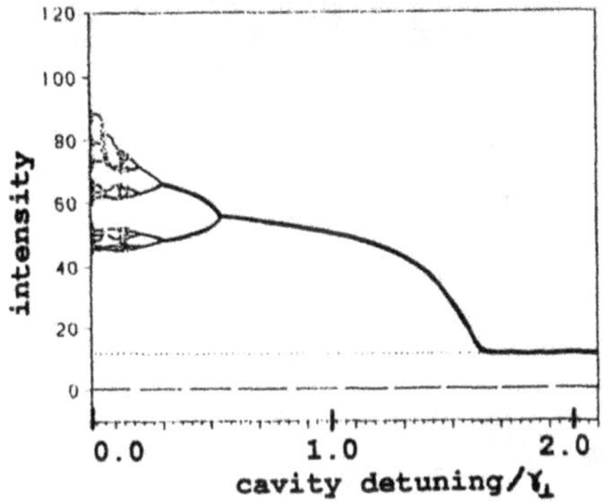

Fig. 14. Bifurcation diagram with peak intensities versus cavity detuning.[23)]

In resonance, it is found that the chaotic solution coexists with the steady state solution in a region below the threshold for the loss of stability of the steady state solution. This means that in this region the laser may behave quiescently or chaotically depending on how it is turned on or on its past history.

It was long believed that this kind of system could not be observed experimentally because the combination of the bad cavity condition and the high level of excitation could not be met. Careful studies by Weiss and Brock[44)] has revealed that a number of optically-pumped transitions offer the opportunity to observe this kind of unstable laser behavior.

Several of their experimental results are presented below. We see that many of the basic features of the model are remarkably well reproduced by the experimental results. The high threshold and the inverse period-doubling with increasing detuning both seem to be in substantial agreement with the predicted behavior.

Careful analysis of optically pumped laser systems indicates the difficulties with making an exact comparison. Since the optical pumping is usually done with another coherent laser, the simple two-level model will not generally apply. Dupertuis and coworkers[45] have analyzed the more general

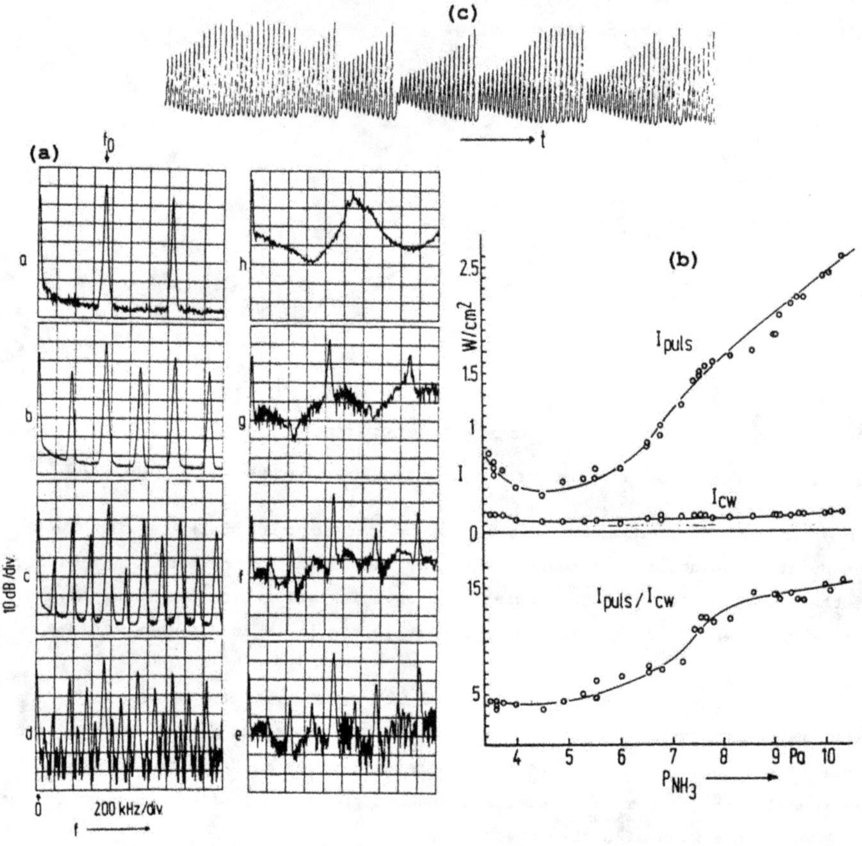

Fig. 15. Results from FIR experiments of Weiss and Brock [44] showing the period-doubling spectra as the laser is tuned to line center (a) and showing that the pulsing threshold in the high pressure regime approaches a value expected from the single-mode, homogeneously-broadened two-level-atom laser theory (b), and a plot of intensity in the chaotic domain (c).

problem and have shown limiting cases in which the two-level theory may be found. These generally require a significant amount of collisional broadening to help remove the additional coherences established by the pump and this may not be easily achieved in the experiments without reducing the possibility of strong pumping, as collisional broadening also dilutes the effectiveness of the excitation.

The more complex model can be solved numerically and compared with the experiments. Meanwhile, the experiments are going forward to provide more specific details of the laser operation. This is a case of interest in achieving a particular form of basic laser behavior driving the experimentalist to create and measure a device which, at present, has no particular technological value. However, we are likely to learn many new things about the behavior of lasers from these studies and many subtle things about field-atom interactions.

5. SUMMARY

Two optical paradigms have been examined as examples of dynamical behavior of field-matter interactions. Periodic behavior of many forms has been observed and it can often be classified according to more general principles of of nonlinear dynamics. Aperiodic behavior that can be classified as chaos has been observed and the characteristics of this chaotic behavior, its response to noise, its difference from stochastic behavior, and its eventual usefulness can now be studied. The field of "Optical Chaos" is now open for further exploration by many specialists in laser physics and particularly by the readers of this volume.

ACKNOWLDEGEMENTS

This work was supported in part by a Research Contract from the U.S. Army Research Office. Helpful conversations with F.T. Arecchi, L.W. Casperson, J.Y. Gao, F.A. Hopf, N. Lawandy, L.A. Lugiato, L.M. Narducci, J.R. Tredicce, and C.O. Weiss are gratefully acknowledged.

REFERENCES

[1] Crutchfield, J.P., Farmer, J.D., Packard, N.H., and Shaw, R.S. "Chaos", Scientific American 255, 46 (1986).

[2] Schaffer, W.M. and Kot, M., "Do Strange Attractors Govern Ecological Systems?", BioScience 35, 342 (1985).

[3] Fisher, A., "Chaos: The Ultimate Asymmetry", Mosaic 17, 24 (1985).

[4] McKean, K., "The Orderly Pursuit of Pure Disorder", Discover 8, 72 (1987).

[5] Edelson, E., "The Ubiquity of Nonlinearity", Mosaic 17, 10 (1986).

[6] Bak, P., "The Devil's Staircase", Physics Today 20, 39 (1986).

[7] Heppenheimer, T.A., "Routes to Chaos", Mosaic 17, 4, (1986).

[8] Cvitanovic, P., Universality in Chaos (Hilger, Bristol, 1984).

[9] Hao, B.-L., Chaos (World Scientific, Singapore, 1984).

[10] Bergé, P., Pomeau, Y., and Vidal, C., Order Within Chaos (Wiley, New York, 1986).

[11] Schuster, H., Deterministic Chaos: An Introduction (Physik-Verlag, Deerfield Beach, 1984).

[12] Holden, A.V., Chaos (Princeton U. Press, Princeton, 1986).

[13] Abraham, R. and Shaw, C.D., Dynamics The Geometry of Behavior (in four volumes) (Aerial Press, Santa Cruz, 1984).

[14] Khaldre, K.Y., and Khokhlov, R.V., Izv. Vyss. Uchebn. Zaved. Radiofiz. 1, 60 (1958); Gurtovnik, A.G., Izv. Vyssh. Uchebn. Zaved. Radiofiz. 1, 83 (1958); Oraevskii, A.N., Radio Eng. Electron. Phys. (USSR) 4, 718 (1959).

[15] Bonifacio, R., and Lugiato, L.A., Lett. Nuovo Cim. 21, 510 (1978).

[16] Ikeda, K., Opt. Commun. 30, 257 (1979); Ikeda, R., Daido, H., and Akimoto, O., Phys. Rev. Lett. 45, 709 (1980); Ikeda, K., and Akimoto, O., Phys. Rev. Lett. 48, 617 (1982); Ikeda, K., Kondo, K., and Akimoto, O., Phys. Rev. Lett. 49, 1467 (1982); Okada, M., and Takizawa, K., IEEE J. Quantum Electron. OE-17, 2135 (1981); Nakatsuka, H., Asaka, S., Itoh, H., Ikeda, K., and Matsuoka, M., Phys. Rev. Lett. 50, 109 (1983).

[17] "Instabilities in Active Optical Media", special issue of J. Opt. Soc. Am. B 2, (1985), edited by N.B. Abraham, L.A. Lugiato, and L.M. Narducci.

[18] Optical Instabilities, edited by R.W. Boyd, M.G. Raymer, and L.M. Narducci (Cambridge U. Press, Cambridge, 1986).

[19] Optical Chaos, edited by J. Chrostowski and N.B. Abraham, Proceedings Vol. 667 (SPIE, Bellingham, 1986).

[20] Ackerhalt, J.R., Milonni, P.W., and Shib, M.-L., Phys. Rep. 128, 205 (1985).

[21] Harrison, R.G., and Biswas, D.J., Prog. Quantum Electron. 11, 127 (1985).

[22] Instabilities and Chaos in Quantum Optics, edited by F.T. Arecchi and R.G. Harrison (Springer-Verlag, Heidelberg, 1987); see also Instabilities and Chaos in Quantum Optics, edited by N.B. Abraham, F.T. Arecchi, and L.A. Lugiato (Plenum, New York, to be published).

[23] Abraham, N.B., Mandel, P., and Narducci, L.M., "Dynamical Instabilities and Pulsations in Lasers" in Progress in Optics, vol XXV, edited by E. Wolf (North Holland, Amsterdam, 1988) (to be published).

[24] Lugiato, L.A., in Progress in Optics, Volume XXI, edited by E. Wolf (North-Holland, Amsterdam, 1984), p. 71; Contemp. Phys. 4, 333 (1983).

[25] Gibbs, H.M., Optical Bistability: Controlling Light with Light (Academic, Orlando, 1985) p. 241.

[26] Gibbs, H.M., Hopf, F.A., Kaplan, D.L., and Shoemaker, R.L., Phys. Rev. Lett. 46, 474 (1981); Hopf, F.A., Kaplan, D.L., Gibbs, H.M., and Shoemaker, R.L., Phys. Rev. A 25, 2172 (1982); Derstine, M.W., Gibbs, H.M., Hopf, F.A., and Kaplan, D.L., Phys. Rev. A 27 3200 (1983).

[27] Gao, J.Y., Yuan, J.M., and Narducci, L.M., Opt. Commun. 44, 201 (1983); Gao, J.Y., Narducci, L.M., Schulman, L.S., Squicciarini, M., and Yuan, J.M., Phys. Rev. A28, 2910 (1983); Narducci, L.M., Bandy, D.K., Gao, J.Y., and Lugiato, L.A., in Synergetics: From Microscopic to Macroscopic, edited by E. Frehland (Springer-Verlag, Berlin, 1984), p. 33.; Gao, J.Y., Narducci, L.M., Sadiky, H., Squicciarini, M., and Yuan, J.M., Phys. Rev. A 30, 901 (1984).

[28] Mandel, P., and Kapral, R., Opt. Commun., 47, 151 (1983); Nardone, P., Mandel, P., and Kapral, R., Phys. Rev. A 33, 2465 (1986).

[29] Vallée, R., and Delisle, C., IEEE J. Quantum Electron. QE-21, 1423 (1985); Vallée, R., and Delisle, C., Phys. Rev. A 34, 309 (1986).

[30] Lugiato, L.A., Asquini, M.L. and Narducci, L.M., Opt. Commun. 41, 450 (1982).

[31] Gao, J.Y., private communication.

[32] Chrostowski, J., Vallée, R., and Delisle, C., Can. J. Phys. 61, 1143

(1983); Jerominek, H., Delisle, C., Pomerleau, J.Y.D., and Tremblay, R., Can. J. Phys. 63, 227 (1985).

[33] Zhang, H.J., Dai, J-H., Yang, J-H., and Gao, C.-X., Opt. Commun. 38, 21 (1981); Song, J.W., Lee, H.Y., Shin, S.Y., and Kwon, Y.S., Appl. Phys. Lett. 43, 14 (1982); Zhang, H.-J., Dai, J.-H., Peng-Ye, W., and Chaio-Ding, J., in Laser Spectroscopy VI, edited by H.P. Weber and W. Luthy (Springer-Verlag, Berlin, 1983), p. 322; Dai, J.-H, Zhang, H.J., Wang, P.-Y., and Jin, C.D., Acta Phys. Sin. 34, 992 (1985).

[34] Orozco, L.A., Rosenberger, A.T., and Kimble, H.J., Phys. Rev. Lett. 53, 2574 (1984); Orozco, L.A., Kimble, H.J., and Rosenberger, A.T., Opt. Commun. 62, 54 (1987).

[35] Macke, B., Segard, B., and Zemmouri, J., Optical Instabilities Workshop, July 1987.

[36] May, R.M., Nature 261, 459 (1976).

[37] Feigenbaum, M.J., J. Stat Phys. 19, 25 (1978); 21, 669 (1979).

[38] Crutchfield, J.P., and Hubermann, B., Phys. Lett. 77A, 407 (1980).

[39] Derstine, M.W., Gibbs, H.M., Hopf, F.A., and Kaplan, D.L., Phys. Rev. A 26, 3720 (1982).

[40] Le Berre, M., Ressayre, E., Tallet, A., and Gibbs, H.M., Phys. Rev. Lett. 56, 274 (1986); Le Berre, M., Ressayre, E., Tallet. A., Gibbs, H.M., Kaplan, D.L., and Rose, M.H., Phys. Rev. A 35, 4020 (1987).

[41] Lu, W., and Tan, W., Opt. Commun. 61, 271 (1987).

[42] Uspenskii, A.V., Radio Eng. Electron. Phys. (USSR) 8, 1145 (1963); 9, 605 (1946); Korobkin, V.V., and Uspenskii, A.V., Sov. Phys. JETP 18, 693 (1964); Graysiuk, A.Z., and Oraevskii, A.N., Radioteckh. Electron. 9, 524 (1964) [Radio Eng. Electron. Phys. (USSR) 9, 424 (1964)]; and in Quantum Electronics and Coherent Light, edited by P.A. Miles (Academic, New York, 1964), p. 192; Haken, H., Z. Phys. 190, 327 (1966); Risken, H., Schmidt, C., and Weidlich, W., Z. Phys. 194, 337 (1966); Graham, R., and Haken, H., Z. Phys. 213, 420 (1968); Risken, H., and Nummedal, K., J. Appl. Phys. 39, 4662 (1968).

[43] Narducci, L.M., Sadiky, H., Lugiato, L.A., and Abraham, N.B., Opt. Commun. 55, 370 (1985).

[44] Weiss, C.O., and Brock, J., Phys. Rev. Lett. 57, 2804 (1986).

[45] Dupertuis, M.A., Salomaa, R.R.E., and Siegrist, M.R., Opt. Commun. 57, 40 (1986).

UNPUBLISHED SEMINARS GIVEN AT THE FIRST INTERNATIONAL COURSE IN NONLINEAR DYNAMICS, MEDELLIN, SEPTEMBER 1-5, 1986

1. Magnetic Surfaces in Tokamaks
   I. L. Caldas

2. Resonant Helical Fields in Tokamacs
   I. L. Caldas

3. Some Criteria of Transition to Global Stochasticity
   J. Mahecha

4. Floquet Theory of Canonical Systems
   E. Piña

5. Channeling in Stability Via Canonical Maps
   A. W. Sáenz

6. Changes in Stability Type of Channeling Orbits
   A. W. Sáenz